T0323641

Social Networks of Meaning and Communication

Social Networks of Meaning and Communication

JAN FUHSE

OXFORD
UNIVERSITY PRESS

OXFORD
UNIVERSITY PRESS

Oxford University Press is a department of the University of Oxford. It furthers
the University's objective of excellence in research, scholarship, and education
by publishing worldwide. Oxford is a registered trade mark of Oxford University
Press in the UK and certain other countries.

Published in the United States of America by Oxford University Press
198 Madison Avenue, New York, NY 10016, United States of America.

Library of Congress Cataloging-in-Publication Data
Names: Fuhse, Jan, 1975– author.
Title: Social networks of meaning and communication / Jan Fuhse.
Description: New York, NY : Oxford University Press, 2022. |
Includes bibliographical references and index.
Identifiers: LCCN 2021024136 (print) | LCCN 2021024137 (ebook) |
ISBN 9780190275433 (hardback) | ISBN 9780197606834 (epub) |
ISBN 9780190275457 (online) | ISBN 9780190275440 (UPDF)
Subjects: LCSH: Social networks. | Interpersonal relations. |
Meaning (Psychology)—Social aspects. | Communication—Social aspects.
Classification: LCC HM741 .F84 2021 (print) |
LCC HM741 (ebook) | DDC 302.3—dc23
LC record available at https://lccn.loc.gov/2021024136
LC ebook record available at https://lccn.loc.gov/2021024137

DOI: 10.1093/oso/9780190275433.001.0001

1 3 5 7 9 8 6 4 2

Printed by Integrated Books International, United States of America

Painting on cover: Milada Marešová, *Wohltätigkeitsbasar*
(*Dobročinný bazar*), 1927, © Milada Maresova Estate

Exhibitor: Kunstforum Ostdeutsche Galerie Regensburg,
on permanent loan of Deutsches Historisches Museum, Berlin

Owner of Mrs Milada Marešová's artwork Copyright and collector:
doc. MUDr. Jan Mareš, CSc. (janmares1@seznam.cz)

Photograph of painting: Lukas & Zink, Fotografen, Regensburg,
on behalf of Kunstforum Ostdeutsche Galerie Regensburg

Contents

Preface

This book is an intermediate stop on my intellectual journey of trying to make sense of informal social structures. In a sense, the journey started in the study for my master's degree at Warwick University in 1998–1999. There I attempted to construct a theory of collective identities, with street gangs and social movements as empirical examples. This interest in collective identities carried into my doctorate at Universität Stuttgart. During my studies of Italian migrants in Germany (Fuhse 2008), I became more and more convinced that networks of social relationships play a major role for their integration, but also in other phenomena delineated by social categories and collective identities, and in social life in general.

My background in Niklas Luhmann's systems theory did not really account for social networks, even with its few "network extensions" available in the early 2000s (see section 1.4.f). The prevalent structuralist perspective in network research left me similarly disappointed, since it simply took its analytical constructs for granted (see chapter 1). I wanted to know what social relationships "are," and why they play such an important part in the social world. In my erratic search for answers, I discovered relational sociology around Harrison White. Like systems theory, the approach considers social structures as constructs of meaning, and as dynamic and only ever fixated provisionally. Maybe this is what social networks really are? I have since become convinced that ontological questions cannot be answered unequivocally, that social reality does not lend itself to conceptual pinpointing once and for all, and that conceptual arguments have their merit in connection to empirical research, if at all. In this regard, relational sociology proves particularly fruitful.

However, I had to make sense of the sprawling and heterogeneous writings of White and his followers by myself, reading in the German province. At the time, I had nobody to talk to about these matters, let alone guide me through the literature. Also, relational sociology does not offer a coherent theoretical system, like Luhmann's systems theory. White's writings consist of theoretical ideas out of, and for, empirical studies that do not always connect. The relational sociologists around him—Peter Bearman, Mustafa Emirbayer,

David Gibson, Roger Gould, Eiko Ikegami, Paul McLean, Ann Mische, John Mohr, John Padgett, Charles Tilly, and many others—only occasionally link their studies and arguments explicitly to White's (Fuhse 2015b).

Hence, I found myself browsing through this body of work, trying hard to find conceptual clarity and consistency. In 2005, I was done doing this alone. I contacted Bearman, Tilly, and White about a possible research visit to Columbia University. They were very welcoming and supported me through my first visit in September 2005, and a later postdoc fellowship in 2007–2008. I learned from them and others that relational sociology is more a way of thinking in relational terms than a coherent theoretical perspective. However, I was, and still am, for theoretical coherence.

At the time, I had already begun linking the ideas of relational sociology with frameworks I knew: systems theory, of course, but also symbolic inter-actionism, Norbert Elias's configurational sociology, and other theoretical approaches. The fruit of this endeavor lies at the core of this book. It offers a theory of social networks as patterns of expectations concerning the (in-formal) behavior of particular actors toward particular other actors. These expectations arise, stabilize, and change over the course of communication. But they also structure communication.

Through this, networks come into interplay with other aspects of the social world, such as social inequalities, categories, formal and spatial organization, social fields, and cultural patterns. At the heart of this interplay lies an under-standing of social relationships and networks as imbued with meaning. This allows extending the perspective to a wider universe of cultural forms, such as group cultures and collective identities, ethnic and gender boundaries, and role categories and institutions.

I have tried to probe the perspective here as far as possible, systematizing research and developing conjectures for future research with regard to the interweaving of social networks with these forms of meaning. However, the book remains focused on informal social structures on the meso-level, even if their interplay with wider culture is examined. Other issues like formal organization, social fields, cultural networks, and communication through media technologies are bracketed here. Also, I do not discuss questions of methods in detail here. All of these extensions are reserved for future re-search, or I have addressed them (provisionally, as always) in papers not in-cluded in the book (Fuhse 2018a, 2020a; Fuhse et al. 2020). Of course, even the topics covered here are not dealt with comprehensively and need much further research, empirical as well as conceptual.

No man (and no woman) is an island, nor a genius (or fool) who emits intellectual arguments out of tangled internal processes alone. The ideas expressed in this book first of all come out of my engagement with the sprawling literature on networks, meaning, and communication. I hope that I have not forgotten too many references that have been important for me. More important, I have benefited from extensive intellectual exchanges with many, many people over the last 15 years. In the following, I only list the most important who have taken their time to read parts of the book manuscript, or of the papers that made it into the book.

John Martin, Wouter de Nooy, and two anonymous reviewers read the full manuscript and gave valuable comments. John and Wouter have been of tremendous help over the last years, not only with this book and with general discussions, as well as with invitations to the University of Chicago and the University of Amsterdam. John supported me with important advice and feedback on the book proposal, too. And importantly, he and Wouter gave me critical feedback on my sprawling and erratic ideas of what to do in the book, saving the readers a lot of trouble.

Ron Breiger, Frédéric Godart, Joscha Legewie, Dean Lusher, Sophie Mützel, Matthias Thiemann, Chuck Tilly, and Harrison White commented on the original paper published in *Sociological Theory* (2009) that morphed into chapter 2 of this book. Harrison and Ron also gave feedback on a lengthy manuscript in an early form, much of which is now contained in chapters 7 and 8 (a shorter version of chapter 8 was published in the *European Journal of Social Theory*, 2015). Both—along with Chuck Tilly and Peter Bearman— have been very supportive of my work and my career and generously hosted me at Columbia University and at the University of Arizona.

I am also grateful for comments and criticisms on this early manuscript for chapters 7 and 8 by Jenni Brichzin, Gwen van Eijk, Ofer Engel, Joe Galaskiewicz, Neha Gondal, Boris Holzer, Corinne Kirschner, Monica Lee, Paul McLean, Ann Mische, John Padgett, Jörg Raab, Marco Schmitt, Eric Schoon, and Hendrik Vollmer, as well as discussions at the 2010 Sunbelt conference, at Rutgers University, at the University of Arizona, and at the University of Chicago. I have to single out Boris for extensive deliberation on the matter, as well as for giving me a job at the University of Bielefeld from 2009 to 2013, when I desperately needed one. Boris, Neha, Sophie, and Marco have been great discussants and coauthors of other pieces over the last years. The arguments developed in collaboration with them have made their way into this book.

Gary Fine and Stefan Kühl provided much-needed feedback on chapter 3. Anna Amelina and Vince Marotta commented on an early version of chapter 4, which was subsequently published in the *Journal of Intercultural Studies* (2012). Vince did a great editorial job for the journal, pushing me gently to a fundamental reorganization of arguments. Chapter 5 was originally published in German in 2012 in the *Berliner Journal für Soziologie* but thoroughly revised here. The comments of Gert Albert, Paul Buckermann, Jens Greve, Richard Heidler, Olaf Kranz, Lena Laube, Gesa Lindemann, Nico Lüdtke, and again Jenni Brichzin and Marco Schmitt helped improve the early version. Chapter 6 benefits from presentations and discussions at the 2019 European Conference on Social Networks in Zürich and at the Mitchell Centre for Social Network Analysis at the University of Manchester in 2019. Alejandro Espinosa gave valuable feedback on the written chapter.

Three chapters have not made it into the book, probably for a more coherent and less sprawling volume. Nevertheless, I want to thank a number of people who read them and helped me with them: Gabriel Abend, Paul Chang, Joe Galaskiewicz, Avid Goldberg, Gene Lerner, Dan McFarland, John Mohr, Paolo Parigi, Kyle Puetz, Craig Rawlings, Geoff Raymond, Jacob Reidhead, Kathia Serrano-Velarde, and again Ron Breiger, Wouter de Nooy, Neha Gondal, Monica Lee, John Levi Martin, Eric Schoon, and the two anonymous reviewers for Oxford University Press. In addition to the people already mentioned, Dan McFarland and John Mohr have been key to discussing methodological issues that now feature only briefly in the book. They kindly hosted me at Stanford University and at the University of California at Santa Barbara and organized discussions of my work there. I take this opportunity to emphasize John's impact on relational sociology, formal studies of culture, computational social science, and on my own thinking on networks of culture, until his untimely death in August 2019.

The Alexander von Humboldt Foundation supported me generously during my postdoc at Columbia University and upon returning to Germany, as well as bestowing research awards to Dan McFarland and John Martin, which were instrumental for our collaboration. The German Research Association (DFG) funded my research from 2013 to 2018 with the even more generous Heisenberg research fellowship. In these years focused entirely on research, most of the book was written. I would like to thank both organizations for their belief in me and my work, especially since my profile and qualifications do not really fit the German academic job market.

Klaus Eder and Steffen Mau organized invaluable institutional support at Humboldt University.

James Cook from Oxford University Press supported the project, helped the manuscript through various stages of disarray, and provided important guidance at crucial points of the publication process. Emily Mackenzie patiently managed the production process at OUP, while Narayanan Srinivasan handled the project at Newgen, Sharmila Munusamy worked on the typesetting, and Maria Cusano competently edited the manuscript. The Deutsches Historisches Museum and the Mareš family generously allowed us to use the wonderful painting by Milada Marešová for the cover. Gabriela Kašková at the Kunsforum Ostdeutsche Galerie in Regensburg was very helpful in organizing the permission and the reproducable photograph. Big thanks to all of them!

Finally, this book would not be possible without the unwavering support of my wife, Lena May, through at times a difficult process, and through the demands on time, nerves, and health of getting children, building a house together, job uncertainty, and the coronavirus pandemic. To her, I dedicate this book.

1

Networks with Theory

What are networks of social relationships? Where do they come from, and what effects do they have? From a structuralist perspective, these questions can be answered in a straightforward way: social networks consist of the *patterns of ties* between actors. These can be represented as lines or arcs in a network graph, or as 1s and 0s in a network matrix. Their effects and their causal antecedents have been studied extensively, and firmly established, in empirical studies: social networks form where people meet (at *foci of activity*; Feld 1981). We connect to people with similar traits, values, and attitudes (homophily; McPherson, Smith-Lovin, and Cook 2001) and with common friends (*transitivity*; Cartwright and Harary 1956), and to those who already have a lot of connections (*preferential attachment*; Barabási and Bonabeau 2003). They help on the *labor market* (Granovetter 1973; Lin, Ensel, and Vaughn 1981), for *upward mobility* in companies (Burt 1992, 115ff), and for *political mobilization* (McAdam 1986; Gould 1995). They accord *power* and *creativity* to people at advantageous network positions (Marsden 1983; Padgett and Ansell 1993; Collins 1998; Burt 2004). Recent advancements in network methods, like exponential random graph models (ERGMs) and the SIENA software, disentangle these various mechanisms and examine their impact on empirically observable network constellations and on individual behavior (Lusher, Koskinen, and Robins 2013; Snijders, van de Bunt, and Steglich 2010).

The structuralist perspective can be summarized as follows (White, Boorman, and Breiger 1976; Wellman 1983; Wellman and Berkowitz 1988; Borgatti and Halgin 2011): Social phenomena are fruitfully studied with regard to the pattern of relations between actors. These patterns of relations—social networks—are the decisive level of social structures. Other features like formal roles, cultural norms and values, or the distribution of attributes across actors are treated as secondary and as mostly determined by networks. The "reality" of social networks is inferred from real observed effects and from their intuitive plausibility.

Social Networks of Meaning and Communication. Jan Fuhse, Oxford University Press. © Oxford University Press 2022.
DOI: 10.1093/oso/9780190275433.003.0001

But structuralism suffers from shortcomings that make it unfit as the un-derlying perspective of network research:

- The insights listed remain mostly unconnected, more of a collection of arguments than a coherent theory.
- Not all supposedly "universal" network mechanisms are at play under all circumstances. For example, preferential attachment may govern citations in science, but not friendship choices in school classes. Friendships tend to transitivity, but love does not.
- We know little about the relations between networks and other aspects of the social world. For example, why are some categories and attributes (ethnicity, age, etc.) conducive to homophily in relatively segregated networks, while others are not (e.g., gender, hair color)?
- Most fundamentally, we do not really know why social networks exist at all, and what role they play in social life. In what sense, and under what circumstances, can we say that social networks are real? And if so, what is their theoretical and epistemological status?

This book aims at going beyond structuralism to provide for a refined theo-retical perspective of network research. This requires confronting network research with theoretical reflection. The book constitutes an endeavor in *re-lational sociology*, the recent push toward grounding the social sciences in "relations," rather than in individuals or holistic systems (Emirbayer 1997; Crossley 2011; Donati 2011; Mische 2011; Dépelteau and Powell 2013; Erikson 2013; Powell and Dépelteau 2013; Prandini 2015; Dépelteau 2018; Fuhse 2020b). This movement is quite heterogeneous and subject to con-flicting definitions and ambitions, as I elaborate in section 1.4. In my defini-tion, "relational sociology theoretically reflects on what social relationships and networks are, and what role they play in the social world" (Fuhse 2018a, 458ff). This specifies the somewhat fuzzy term of "relations" as "social relationships," and it firmly connects "relational sociology" to network re-search. Network research is in need of theoretical reflection and grounding, and theory always needs to connect to empirical research. But of course, other authors understand relational sociology differently, and I do not mean to dismiss their efforts here.

Now, there is currently a rich and important debate in American soci-ology about "theory" and "theorizing" (Abend 2008; Reed 2011; Swedberg 2017). Theory consists of generalizations of empirical observations in the form of abstract concepts and connections between them (see section 1.3). Empirical observations are themselves theory laden, as philosophy of science teaches us (Quine 1951; Hanson 1958; Feyerabend 1962; Hesse 1962, 15ff). We cannot and do not really start our observations without theory and then distill concepts and the links between them by generalizing our observations. Rather, our observations are already informed and focused by theoretical frameworks, be they acknowledged and explicitly formulated or unreflected and unacknowledged.

Consequently, we do not construct new theory from empirical observations, but from confronting already existing theory with them. This is what I do in this book: Drawing on the rich repertoire of theoretical approaches in sociology and in related disciplines, I combine a number of diverse concepts and fit them to empirical observations of network constellations. This follows Arthur Stinchcombe's concept of "crafting" theories in interchange with empirical material, rather than deriving them deductively from abstract principles (2005, 296ff). Some of these observations come from network research, and others from specialized areas of inquiry such as the sociology and anthropology of ethnic groups. If the outcome of these explorations is original, it is not because I invent something entirely new, to come up with idiosyncratic vocabulary like Erving Goffman or Harrison White. Rather, I combine existing concepts with a clear and well-developed meaning like network, communication, culture, category, institution, and role. Hopefully, this helps in rendering the resulting theory as clear and well developed as possible.

1.1. Foundations

Where do these combinations lead? What kind of conceptual architecture awaits the reader in the chapters to come? I will flesh out the theoretical perspective in chapters 2, 5, and 8 and connect it to empirical research in the other chapters. Here I offer a first look at the trinity of network, meaning, and communication, as the key terms on which this perspective rests.

I start from the intuition that *social networks as clear-cut, seemingly unproblematic and stable patterns do not exist.* Social relationships are not readily reducible to 1s (existent) and 0s (nonexistent), and they are often hard to discern and confusing for the actors themselves. Also, no two relationships are alike, though network research relies on the similarity of existing (and of nonexisting) ties in a network. And relationships continuously change over time, with extensive periods of inactivity where the status of relationships remains in limbo.

So why do we still talk about social relationships and networks, and why do they stubbornly show effects in empirical studies? To the extent that social relationships and networks exist as relatively durable structures of the social world (and not just as methodological constructs), I propose that they consist of two interrelated aspects: (1) observable regularities in communication—how often and how do actors communicate with each other?—and (2) patterns of meaning developing over the course of communication, and stabilizing it in turn. In short, communication leads to relationships and networks as structures of meaning, and these make for observable regularities in communication. Let me sketch these two steps in more detail.

First, in line with recent scholarship on the intersection of culture and networks, I conceptualize social networks as symbolic constructions, as patterns of *meaning* (Fine and Kleinman 1983; White 1992, 2008; Emirbayer and Goodwin 1994; Mische 2003, 2011; Crossley 2011; Pachucki and Breiger 2010; McLean 2017). In my formulation, dyadic social relationships are bundles of *expectations* about the behavior of alter and ego toward each other (chapter 2). Relationships and networks can be termed "relational definitions of the situation" (following Thomas, Merton, and Goffman) governing the course of communication. These relational expectations or definitions are connected to other forms of meaning: the identities of the actors involved, social categories with prescriptions for how to interact within and between categories (e.g., gender), institutionalized role categories (e.g., professor/student), and cultural models for relationships (love, friendship, patronage, etc.). All of these prescribe particular kinds of network constellations, and these stabilize if the communication in them follows these social categories, roles, or cultural models.

Now, the exact meaning of "meaning" is difficult to pin down. As I understand it, "meaning" does not denote a delimited social phenomenon but, rather, a particular aspect of all social life (see section 2.2). It covers what comes into view if we adopt an interpretive stance and ask what social phenomena *mean*, in the methodological move that Max Weber calls "understanding" ([1913] 1981). For example, we can look at "poverty" and measure it structurally in terms of quantitative criteria. Or we can ask for the meaning of poverty: How do people understand their own poverty, or that of others? And how is poverty invoked in communication—with social categories, through particular linguistic forms, as a lack of ways for expressing things? Similarly, we can examine networks structurally as the pattern of links between nodes. Or we can ask for the meaning of social relationships and networks: How do people see each other, and their relationships to each other? How do networks feature in communication? Of course, the theory claims that the structure and meaning of social networks are intertwined. But we separate the two conceptually to examine this intertwining, for example, by examining how linguistic forms are used to negotiate and define social relationships.

In this sense, "meaning" is an umbrella term covering a wide range of more clearly defined concepts: identities, expectations, roles, categories, etc. Like Weber, we can locate meaning subjectively in the heads of the individual actors involved: How do people understand their situation, the social structures they are embedded in? Or we can conceptualize meaning as processed in communication—in the linguistic forms, the definitions of the situation, and the categories and roles invoked in communication. I opt for the second option in chapters 7 and 8. Most of the arguments from chapters 2 and 6 are compatible with either subjective or communicative meaning. Generally, meaning is processed both in people's heads and in the communication between them. But it seems prudent to focus on one of the two conceptually and methodologically.

The second step leads away from static imagery to a dynamic reconceptualization of relationships and networks. Expectations form, reproduce, and change over the course of *communicative events*. These events—and not the relations arising from them—are the basic stuff of the social world. This makes for a thoroughly processual vision, as laid out by Alfred North Whitehead ([1929] 1978) as opposed to the relational ontology of Ernst Cassirer ([1910] 1923)—frequently heralded as a forbear of relational sociology. Social relationships exist not as basic building blocks, but as accomplished constructions, insofar as they form and reproduce over the course of events.

Such events have been termed "transactions" in relational sociology by Mustafa Emirbayer (1997, 287) and Charles Tilly (2005a, 6f). Other authors conceptualize the basic processes in networks as "action," "interaction," or "exchange" (Cook and Emerson 1978; Burt 1982; Fine and Kleinman 1983; Hedström 2005; Crossley 2011; see chapter 7). Most of these authors view social networks as based on the subjective processing of meaning by the individuals involved. In contrast, I draw on Niklas Luhmann's theory of communication to argue that social networks are fruitfully conceptualized as patterns of relational expectations (relational definitions of the situation) emerging, stabilizing, and changing over the course of communicative events (see chapter 8). The process of communication entails the attribution of events to actors with specific dispositions, which leads to expectations about their behavior to others. Such communicatively constructed relational expectations constitute the meaning structure of social networks and make for observable regularities of communication.

These two steps lead to uneven triad of the three key terms. The basic process of everything social is *communication*, conceptualized here as the suprapersonal processing of meaning in communicative events (Watzlawick, Beavin, and Jackson 1967; Luhmann [1984] 1995, 137ff; 2002, 155ff). These events are often language based, but they need not be. Some of the most important features of communicative events are non-linguistic, from gestures and facial expressions to handshakes, hugs, kisses, and even physical violence. Many incidents thus qualify as communicative events, as long as they convey information and relate actors. Communicative events leave a trace in the social world not through the intentions, knowledge, or other subjective meaning of actors, but through their being understood and reacted upon. Michel Foucault bases his discourse analysis on a similar idea of utterances ("énoncés") relating to each other ([1969] 1972). And conversation analysis views interactional turns as produced, and as determined in their meaning by their conversational environment, rather than by autonomous subjects (Schegloff 2007, xiv).

Since the trace of communicative events is tied to their being seen, and understood in a certain way, it lies in their *meaning*. In this perspective, communicative events, like texts or documents, carry a meaning relatively independently of the intentions of their authors or speakers. While an utterance can be subjectively meant in one way, it may be understood quite differently. Both the sender's intentions and the receivers' interpretations remain unknown—not only to observers, but also to other participants in communication. All of us have to rely on what is communicated, rather than what is thought. Consequently, the meaning of a communicative event—whether an incidence constitutes joke or insult, casual remark or invitation—is only realized over the course of subsequent events (Schneider 2000).

All social structures and cultural patterns result from this processing of meaning in communicative events, and from their traces crystallizing in symbols (linguistic and non-linguistic), norms, institutions, and social categories; in collective identities (like street gangs and social movements); in companies, universities, and state administrations; but also in fields of society like the economy, politics, law, or science. All of these have to be enacted in communicative events, and they constantly change through the course of communication. Following Max Weber and Talcott Parsons, we can say that all social structures consist of expectations (Weber [1913] 1981, 159ff; Parsons et al. [1951] 1959). These link communicative events to each other (Luhmann [1984] 1995, 96f). Every event builds on the expectations formulated and established in previous communication, and it reproduces or changes them to channel subsequent communication.

Social networks are not a master concept for these diverse social structures, but a special case. They certainly play an important role for norms and institutions, collective identities, culture in general, formal organizations, and social fields. But these phenomena should not be reduced to networks. I argue in chapter 8 that social relationships and networks form one particular type of social expectation, with other social and cultural structures made of other kinds of expectations. One of the main tasks of this book is to flesh out: How exactly do social networks come about in communication? Which aspects of communication lead to the expectations embodied in networks? And what is specific about these networked expectations as opposed to other kinds of social structures?

Briefly put, social networks consist of social relationships in which actors behave toward each other in particular ways. Such observable *communicative regularities* are tied to *relational expectations* about how the actors are supposed to behave toward each other. Relational expectations are established in communication when communicative events are interpreted as based on dispositions of actors toward other actors. But they rarely start from scratch, instead building on cultural models for how to relate between lovers, friends, patron and client, and family members, and within and between social categories like gender and ethnicity.

This notion of social networks diverges from the purely analytical one prevalent in social network analysis. We move here from using networks as a methodological construct to treating them as social phenomena, asking what lies behind the methodological construct. Also, we move away from "network" as a metaphor used for a wide array of phenomena, as, for example, in recent formulations of a "network society" (Castells 2010; van Dijk 2012). If we prefer rigorous theorizing to fancy toying with words, we need to get our concepts right. This entails clearly defining them, and delineating what phenomena (and what aspects of them) the concept at hand refers to.

The remainder of this introductory chapter addresses a few preliminaries. Do we actually need theory in network research (section 1.2)? Unsurprisingly, my answer is yes. And what kind of theory do we need (section 1.3)? I argue that, ideally, sociological theories are coherent perspectives that allow us to examine certain aspects of an overly complex social reality. They are devices for conducting and interpreting empirical research, and for putting its results into a wider context. This means that theories can never provide an ultimately correct view of social reality. Section 1.4 then gives an overview of the most important current theoretical approaches to social networks and "relational sociologies," flanking and rivaling my own endeavor. A brief outline of the book with short summaries of the chapters follows (section 1.5). Of course, the impatient reader may skip all of this and jump right to the theoretical core in chapter 2.

1.2. Why theory?

Do we really need theory? Does anybody doing network research on social constellations have to read the sprawling reflections in this book? Network scholars often seem to think that we can do without theory. Thereby they adhere not to *theory abstention*, but to naive, lay conceptions of the social world. This includes the idea that network mechanisms like reciprocity, transitivity, or preferential attachment are as much theory as we need. But theory is also involved in our research methods and in our empirical observations (Quine 1951; Hanson 1958; Feyerabend 1962). If we measure networks of social relationships, we follow the ideas that (a) social relationships are discernible things, and (b) that their patterns matter. These ideas are theoretical insofar as they already form part of a logically integrated framework (e.g., the ones offered by Azarian 2010; Crossley 2011; or White 2008). However, much network analysis does not overly care about these theoretical underpinnings. So the ideas often remain proto-theoretical—unfounded and under-reflected. This means that many studies do not rest on the careful consideration of whether the social relations observed actually exist, in what way, and how their patterns should actually make a difference.

To be fair, all empirically oriented branches of social science seem to breed theory abstention, from number crunching in statistical survey research to qualitative research that deliberately stays away from theorizing in order to arrive at "non-distorted" and comprehensive pictures of their objects of study. None of these are devoid of theory. They build on theoretical ideas encapsulated in their methods, and/or they follow under-reflected lay notions of the social world. I hold the widespread attitude of theory abstention to be ill-conceived, in network research and elsewhere.

Ideally, a refined theoretical perspective would improve network research in four ways:

- First, it would provide a more "realistic" and plausible *conceptualization* of actual social networks. Theory provides a coherent language for talking about the social world.
- Second, this would give us a better sense of what the prevalent methods of detecting and analyzing networks do, and what their results mean. In other words, theory can improve the *interpretation* of our findings. For example, the theory developed here suggests that people do not have a well-defined set of personal relationships around them, and that the current discussion of a changing size of these personal networks is ill-conceived (McPherson, Smith-Lovin, and Brashears 2006).
- Third, theory should yield nontrivial *expectations* about what to find in empirical research. These conjectures or hypotheses would have to be tested, which could lead to the confirmation or refutation of parts of the theory, or to its modification and elaboration.
- Finally, theory points to certain aspects of the social world as important. This leads to *methods* to observe these features, which we could then study in their connections to other aspects of the social world. The theory proposed here suggests that we should study the definitions of identities and relationships in networks (their meaning structure; chapter 2), and their negotiation in ongoing communication (chapter 8).

Overall, then, theoretical reflection should lead to better empirical research and/or to improvement in our interpretation. Of course, this means that theory should be intimately connected to empirical observations and to our methods for arriving at them. Unfortunately, the widespread theory abstention of empirically minded scholars is flanked by *theoretical discourses* that are mostly *self-centered* and far removed from the problems of empirical research. Many publications discuss classical sociological theories (Marx, Durkheim, Simmel, Weber, Dewey, Mead, Goffman, Bourdieu, Habermas) or focus on social theoretical notions like emergence or individual agency (Abend 2008, 179, 181). Often enough, they toy with concepts with regard to their normative or philosophical underpinnings, but without as much as a side glance at application to empirical social phenomena. This divorce of theory from empirical orientation feeds back into the theory abstention of empirical scholars. Theoretical discussions of empirical results are criticized by reviewers and colleagues for the disregard (or short-handed discussion) of Durkheim or Weber, and for a lack of attention to human agency or to the problem of emergence. In reaction, many empirical sociologists stay mum about theory altogether.

1.3. What theory?

This double bind of abstention from, and self-centeredness of, theory is not easily dissolved. For the theory side, I suggest that we should construct frameworks that are connected to and conducive to empirical research. Theories are packaged sets of sentences about the world (Popper [1935] 2002, 37f, 50f; Quine 1951, 39ff; Hesse 1974, 17ff; Fuchs 2001a, 251ff). These sentences connect concepts to each other in meaningful ways, making for a *logically integrated perspective* that effectively restrains what can be said. A theory is not adjustable at will—it comes with rigid connections between its concepts and with clear-cut expectations about empirical observations. As long as these conform broadly to the theory, it counts as "empirically adequate"—according to Bas van Fraassen the best fit of theory to the empirical world that we can hope for (1980).

Following Ronald Giere, I view theories as "perspectives" that we adopt to make sense of empirical phenomena (2006a). Any theory focuses on particular aspects of the phenomenon at hand, rendering them clearly visible, while ignoring others. Of course, multiple perspectives on the same phenomenon are possible, without an objective yardstick to determine which one is better (Giere 2006b). Accordingly, I do not strive to present the objectively correct theory of social networks, but one that is internally coherent and leads to interesting observations and nontrivial insights.

Karl Popper classifies the sentences of a theory into assumptions, definitions (or in a broader sense: bridging sentences), and conjectures ([1935] 2002). These are of very different epistemological character and have to be assessed accordingly:

- *Assumptions* (or axioms) are the abstract foundations on which every theory rests. They can never be true or false, since they do not lend themselves to empirical tests. For example, rational choice builds on the idea that human individuals always pursue the course of action promising them the best results. In itself, this idea cannot be tested—it just makes for a starting point for additional reasoning.
- *Definitions* relate concepts to each other (within the theory at hand). Like assumptions, they cannot be right or wrong, but they do focus our attention to particular aspects of the phenomena at hand. Different theories frequently offer rival definitions of the same terms.
- *Conjectures* (or hypotheses) formulate the theoretical expectations to test in empirical research. They can turn out to be true versus false, or corroborated versus refuted. In a way, theories are geared at producing and integrating conjectures that are supported by empirical evidence. However, conjectures refuted by empirical evidence can always be replaced by revised conjectures within the same theoretical perspective. Also, different conjectures can be formulated and corroborated concerning the same phenomena.

This three-fold classification builds on David Hume's distinction between scientific statements that are internal to the theory ("analytic"), like assumptions, and those that refer to empirical observations ("synthetic"), like conjectures. The philosophy of science has dismissed this distinction as untenable, since all concepts and statements contain theoretical assumptions as well as empirical referents (Quine 1951; Feyerabend 1962). Still, I find Popper's triad a useful heuristic tool. It helps us see which theoretical statements serve mainly to integrate and specify theory (assumptions and definitions), and which ones (conjectures) can and should correspond to empirical observations (which themselves rely on theoretical notions).

For Popper, the classification is part of his hypothetico-deductive model where assumptions should lead to conjectures and then be refuted or corroborated on the basis of empirical tests ([1963] 2002). I do not share his insistence on deduction—theoretical arguments can spring deductively from theoretical assumptions, or they can develop abductively from the abstraction and generalization of empirical observations. In line with the recent discussion on "theorizing" (Swedberg 2014, 2017) and with the older emphasis on "theory construction" (Stinchcombe 1968), theories are continuously "under construction." Following Stinchcombe, they are "crafted" in the exchange between empirical observation and theoretical abstraction (as discussed earlier). Otherwise, theory runs the risk of purely metaphysical speculations without connection to the empirical world (and to our methods for observing it).

I follow this general model in the development of the theoretical perspective in this book. The methods and diverse insights from network research are confronted with established sociological theories. In particular, I draw on Norbert Elias's configurational sociology and Harrison White's theory of networks, on theories of meaning like symbolic interactionism and neo-institutionalism, on Erving Goffman's dramaturgical approach, and on the theory of communication by Niklas Luhmann and others. On the empirical side, diverse studies, in particular from relational sociology (around White), provide insights into the interplay of (various forms of) meaning and social networks.

Ideally, this confrontation leads to a theory that:

(1) is as lean and straightforward as possible,

(2) is geared at providing a coherent framework for interpreting and devising network research (both in terms of methods and theoretical expectations) as well as being informed by it,

(3) provides clear-cut definitions of core concepts and connects them to each other,

(4) stays away from ontological assumptions and from inferring unobservable processes without empirical correlates, and

(5) refrains from broad statements that cannot be studied empirically.

With regard to scope, the theory is as abstract and general as possible. However, it remains confined to the interplay of micro-processes of communication with the meso-level of sociocultural configurations. Phenomena to be covered include social relationships and groups, inter-ethnic constellations, love and gender, and collective actors (street gangs and social movements). In this book, I avoid the macro-level of societal structures and broad cultural trends. We have to engage with the question of how social networks are patterned by grand social structures (in particular, fields of society like the economy, politics, or science) elsewhere (Powell et al. 2005; Padgett and Powell 2012a; Erikson and Occhiuto 2017; Fuhse 2020a).

1.4. Theories of networks and relational sociologies

Mine is not the first attempt at formulating a theory of social networks. By now, we face a wide variety of network theorizing, and a number of self-proclaimed "relational sociologies." In chapter 7, I survey how different concepts for social events have been (or could be) connected to social networks: behavior, action, practices, exchange, interaction, etc. I offer overviews of the most important contemporary theories of social networks and of White's relational sociology elsewhere (Fuhse 2015b, 2020c). In this section, I briefly discuss how my theoretical account differs from a number of currently prominent approaches to social networks, and how it picks up on them. The primary focus will be various approaches stylized as "relational sociology" by its proponents. But I also briefly comment on actor–network theory (ANT), on Bourdieu's theory of fields, and on systems theoretical thinking on networks. Given that I regard theories as perspectives, I cannot dismiss these rival approaches as long as they offer coherent visions of the social world. Rather, I focus on how their visions differ from mine, and where I see my perspective as more fruitful than theirs. The emphasis lies on differences, in spite of the many affinities and similarities in our endeavors.

(a) Pragmatist and interactionist approaches

A number of prominent authors like Mustafa Emirbayer (1997; Emirbayer and Goodwin 1994), Nick Crossley (2011), and John Levi Martin (2009, 2011) follow pragmatism and symbolic interactionism in their conceptualizations of social networks (Fuhse 2020c, 38ff; cf. Fine and Kleinman 1983; Erikson 2013; section 7.6). They centrally build on the subjective processing of meaning, in interplay with its negotiation in the interactive process. The double focus on meaning and on process resembles very much my own. However, given the unobservable nature of subjective meaning, the pragmatist and interactionist can often only claim that networks are reflected in the minds of individuals, and affected by their creative reasoning. We do not really have research methods tackling this. But qualitative interviews and ethnographic methods can get us closer to this multifaceted picture of individuals enmeshed in networks in thought and interaction (Crossley 2010; Desmond 2014).

Crossley offers perhaps the most coherent theory of social networks as constituted in symbolic interaction (2011, 28ff). Following Merleau-Ponty, he sees social relationships forming in the intense back and forth between two actors in face-to-face interaction, leading to the exchange of symbolic forms and to a commonality of perspective. Such worldviews and cultural forms diffuse in network structures, making for distinct cultural styles, such as those of the British punk and post-punk movements (2015b). As a consequence, Crossley's theory is mostly about symmetric and cooperative personal relationships and does not dwell much on asymmetric ties and role patterns.

Building mostly on American pragmatism, *Emirbayer* stresses the importance of culture and of individual agency in social networks (Emirbayer and Goodwin 1994; Emirbayer and Mische 1998). However, he draws on John Dewey and Arthur Bentley's concept of "transactions" as foundational for relational sociology, without elaborating how his twin concerns for culture and agency square with this concept (Emirbayer 1997; see section 7.7). In later works, he adopts Pierre Bourdieu's theory of fields, arguing for a stronger concern for "objective relations" in relative distributions of different types of capital (including "racial capital"; Emirbayer and Desmond 2015; Emirbayer and Johnson 2008). Overall, this does not exactly add up to a coherent conceptual universe. And social networks no longer lie at the core of Emirbayer's work, with other features becoming more important: fields and capitals, but also socioemotional constellations, and so forth.

Martin pursues a different strategy by tracing network processes to relatively simple heuristics tied to positions in a network (2009, 2011). While he ties this endeavor to pragmatism and the notion of interaction, it could also feature as a theory of action. Again, Martin's perspective is connected with field thinking (2003). But he relies less on Bourdieu than on German Gestalt psychology.

In contrast to Crossley, Emirbayer, and Martin, I place a stronger emphasis on the inner dynamics of communication. The pragmatist and interactionist approaches remain wedded to the idea that we should account for behavior and for social structures referring to subjective states and processes. As I argue in chapter 7, this focus does not really help us in elucidating social structures and processes, as long as the subjective meaning of individuals remains obscure to us.

(b) Relational sociology around Harrison White

Though intertwined, I distinguish relational sociology around Harrison White (including my own approach) from the pragmatist and interactionist approaches by its reluctance to embrace individual subjects and their subjective processing of meaning. This has been criticized by Emirbayer and Goodwin (1994, 1437f), Christian Smith (2010, 251ff), and others, mostly on ontological and moral grounds. But it makes for a thorough focus on observable aspects of social structure and process. Relational sociology conceptualizes social networks as structures of meaning that are enacted and negotiated in communication (Pachucki and Breiger 2010; Mische 2011; Fuhse 2015b). This happens in the stories told about identities (White 1992, 65ff; Tilly 2002, 26f), in "switchings" between different sociocultural contexts (White 1995a; Mische and White 1998; Godart and White 2010), and in the deployment of cultural frames like friendship, family, or patronage to define and negotiate relationships (McLean 1998). I build on these authors and consider myself as adhering to their approach. My account differs from theirs in two regards:

(1) Social relationships and networks are certainly symbolically constructed and negotiated between the actors at play. But their structures of meaning go beyond stories told. I use the wider term "relational expectations" for how actors should and will behave toward each other. Storytelling forms an important vehicle for explicating and representing these expectations. But communication leads to relational expectations even in the absence of stories (see sections 2.4 and 8.6). This connects well with other forms of meaning affecting the pattering of social networks: social categories and roles, and models for relationships and for actorhood (see section 2.6, chapters 4 and 5). These constitute "relational institutions" that provide cultural models for relationships and network structures. Once adopted in communication, they come with particular expectations how (role-typed) actors will behave toward each other.

(2) I centrally focus on networks as *processed in communication*. White starts in 1992 with a conceptualization of networks as more or less stable structures. Later on, he incorporates "switchings" as responsible for change in sociocultural structures (1995a; Godart and White

2010), and he picks up on Luhmann's concept of communication (with me as one coauthor; White et al. 2007). Overall, however, he went only halfway to a processual notion of networks. Other authors like Ann Mische (2003), Charles Tilly (2005a), Sophie Mützel (2009), and John Padgett (2012a) go further. But they do not really offer a full-fledged conceptualization of networks as processed in communication.

Within relational sociology, Eric Leifer, Paul McLean, Daniel McFarland, David Gibson, and Wouter de Nooy empirically study social networks in communicative processes. They examine either the impact of social relationships on communicative events (McFarland 2001, 2004; Gibson 2005), the effects of events on relationships (McFarland, Jurafsky, and Rawlings 2013), or the ongoing negotiation of roles and relations in communication (Leifer 1988; McLean 1998, 2007; de Nooy 2011, 2015; de Nooy and Kleinnijenhuis 2013). Up to now, these endeavors are only loosely connected to the theoretical approach of White and others. I hope to offer theoretical integration that could lead to methods for studying social relations in communication (chapter 8).

(c) Relational work

Mainly connected to Tilly, the relational work approach by Viviana Zelizer can be considered a close cousin to relational sociology. It examines how people negotiate and demarcate their personal relationships with reference to frames like friendship, love, companionship, etc. (Zelizer 2005; Bandelj 2012). This strongly resembles the work by McLean on how Florentine patricians invoke particular vocabularies of relationships (family, patronage) in their networking attempts (1998, 2007). I incorporate both in my theoretical account of the construction of the meaning of social relationships (their "definition of the relationships") in communication (chapter 2). This allows connecting network research conceptually with the more qualitative stance of McLean and of the relational work approach. I view qualitative research techniques not opposed to formal network analysis and statistical analyses of egocentric networks, but complementing them (Fuhse and Mützel 2011). Following the work of McLean, McFarland, and Gibson, these qualitative research techniques focus on the communicative processing of meaning (as in sociolinguistics and conversation analysis), rather than on the subjective meaning of individuals (as in qualitative interviews).

(d) Other relational sociologies

Apart from the work around White, a number of other approaches are featured under the fashionable label of "relational sociology." Emirbayer's famous article "Manifesto for a Relational Sociology" covered both White's work (along with Padgett's, Tilly's, etc.) and the pragmatist and interactionist approaches favored by himself (1997). For Emirbayer, the label comprises approaches that are neither individualist nor holistic but focus on relations (as opposed to essences). Apart from Georg Simmel, Ernst Cassirer, Norbert Elias, and Ronald Burt as obvious candidates, he christens Karl Marx, Michel Foucault, Pierre Bourdieu, Randall Collins, and partly even Emile Durkheim and Talcott Parsons as relational sociologists. Thinking in relations rather than essences seems to be not only important, but also pervasive, almost ubiquitous in social theory. A similarly wide understanding of "relational sociology" is adopted in the two volumes edited by François Dépelteau and Christopher Powell (2013; Powell and Dépelteau 2013) and in the *Palgrave Handbook of Relational Sociology* (Dépelteau 2018; Fuhse 2020b).

"Relational sociology" was first proclaimed by Italian sociologist *Pierpaolo Donati* (1983). He argues that social relations constitute a reality sui generis, that society is composed of relations, and that sociology should take relations as ontological and epistemological starting points (2011). Donati defines social relations as the "emergent effect" of the actions of multiple actors (2013, 115f). This is not far from the conceptualization offered here. But Donati does not specify how actions (or communications) should have these effects (through the expectations they establish, as I argue). Also, he does not connect to social network analysis. His social relations comprise dyadic friendships and couples as well as informal groupings like the family, voluntary associations, and local communities.

Donati collaborates with Margaret Archer in exploring the "morphogenetic" development of social relations, and the morphogenesis of society through relations (Donati and Archer 2015). Their work builds on the epistemological position of critical realism (following Roy Bhaskar; see also Donati 2015). Critical realism opposes constructivism and postmodernism, but also Karl Popper's critical rationalism. In the social sciences, it postulates that an objective reality exists, and it uses theoretical-philosophical inspection to outline it (Cruickshank 2010; Reed 2011, 56ff). Consequently, Donati and Archer fault the theory around White, and the pragmatist and interactionist approaches for being overly constructivist and for the simplification and reduction of the complex reality of social relations (Donati and Archer 2015, 8ff, 19ff). In contrast to Donati and Archer, I hold that not only social reality is constructed, but also our theoretical representations of it. Hence, simplifications and reductions are unavoidable, and necessary to arrive at theoretical models that are straightforward, manageable, and useful in empirical research (see section 1.3). Also, I accord a lesser role to individual intentions, emotions, and agency. While they might ontologically *be* there and matter for social relationships, they remain elusive and hard to observe. Therefore, we should not pretend to know about them (Fuchs 2001b).

An opposite position to Donati's is taken by Canadian sociologist *François Dépelteau* (2008, 2015).[1] Building on Dewey and Bentley, and Emirbayer, Dépelteau views "transactions" as the basic units of the social world (see section 7.7). These make for a "flat ontology," without relationships as enduring social structures or individuals as privileged cornerstones. Past transactions make for the constellations that current transactions build on. And they lead to the formation of "habits" of actors (Dépelteau 2008, 60f). Like Dépelteau, I argue for thinking about the social in terms of processes, not static structures. However, for me, past process affects the present through the buildup of expectations in relationships, formal organizations, cultural patterns, and fields of society. Where Donati emphasized relational structure, Dépelteau denies any form of social structure, with the exception of ephemeral constellations and individual habits (which he does not really dwell on). Since he also introduces "nonhuman transactors" and focuses on the interplay between researchers and observed phenomena, Dépelteau moves closer to ANT than to other versions of relational sociology (2015).

Most other self-proclaimed relational sociologists fall closer to Donati than to Dépelteau. They see social relations as the central units of the social world, sometimes even as the only one (Dépelteau and Powell 2013; see in particular Powell 2013). I do not share this euphoria around social relations—there are many processes in the social world that are not organized around relationships. For example, the modern economy features transactions between buyers and sellers that are governed by prizes, brands,

---

[1] Dépelteau was key in organizing the recent social theoretical discussions and a number of publications on "relational sociology." This includes the books edited with Christopher Powell in 2013, the *Palgrave Handbook* (Dépelteau 2018), the book series "Palgrave Studies in Relational Sociology," and the research cluster "relational sociology" in the Canadian Sociological Association. He died untimely in August 2018 from cancer. It remains to be seen whether and how the discourse on relational sociology continues. See Dépelteau et al. (2015) for an extended discussion of the possible meanings of "relational sociology" between him, a few others, and myself.

and public information, rather than by relationships between the particular actors involved. Also, people often share ideas and information without communicating directly—for example, through the mass media.

Social relations in relational sociology often seem to cover just about every feature of the social world, and even the non-social. But if we include relations between classes (Marx, Bourdieu), between symbolic forms (Foucault), and between individual and material transactors (Latour, Callon, Dépelteau), the notion of relations becomes pointless. For example, *Pierre Bourdieu* argues for "relational thinking" ([1980] 1990), and he frequently appears as a key reference author of relational sociology (Emirbayer and Desmond 2015; Singh 2016; Papilloud and Schultze 2018). However, his theory accords little importance to actual social relationships, instead focusing on relations of "more or less" of various forms of capital (economic, cultural, symbolic) and on symbolic oppositions between practices (Bottero 2009; Mohr 2013). If differences in symbolic practices and having more or less individual resources constitute relations, then all statistical analyses in empirical social research, and all reconstruction of patterns of meaning in qualitative sociology, qualify as relational sociology. Why not just call this "sociology"?

My own endeavor is more modest. I hope to shed light only on the development of social relationships, on their role in the social world, and on their meshing in social networks. This reduced focus helps making sense of relatively confined social phenomena, while remaining blind to other aspects. This is a much-reduced project for relational sociology, but one that leads to clearer concepts and connects well with empirical research on social networks.

(e) Actor–network theory

Similar arguments apply to ANT around Bruno Latour, Michel Callon, Annemarie Mol, and John Law. Coming from studies of science and technology, ANT argues that we should describe arrangements of human actors and material objects in terms of associations between them (Latour 2005). Both kinds of "actants" (human and material) acquire their meaning only in processes between them, and scientific findings result from action performed by these networks, rather than from individual actors. The constructivist stance and the focus on processes between entities resembles those of relational sociology, as Mützel notes (2009).

However, ANT formulates a principle of symmetry (Latour 1993, 94ff): human and nonhuman actors are to be treated equally by the researcher, with the same concepts applied to them. Following this principle, we have to conceptualize the processes and relations ("associations") between humans, between human and nonhuman actors, and between nonhuman actors in the same way. Apart from material objects, ANT extends these arguments to words systematically co-occurring in texts (Callon et al. 1983) and to scientific keywords that are linked in databases (Latour et al. 2012).

I would argue that these different kinds of relations develop and proceed differently: Between human actors, "relational expectations" evolve as to how these particular actors behave toward each other. This makes for my friends treating me differently from other people. The same holds for collective and corporate actors, both dismissed by Latour. With regard to material objects, some people undoubtedly acquire an attachment to some of them. But these would probably still react to the same stimuli coming from different people indiscriminately. And not many people would claim that similar attachments form between nonmaterial objects, such that one billiard ball reacts differently to being hit by a "significant other" ball. The principle of symmetry forbids us to qualify these different processes and kinds of relations, and therefore to say anything meaningful about the relationships formed between individuals and between collective and corporate actors. Probably in reaction to these problems, Latour introduces different "modes of existence" or ways of relating between entities (2012). This amounts to a "phenomenological" catalog of meanings associated with relations. But it does not offer a substantial account of what happens in associations.

(f) Systems theory

Finally, my thinking has developed in tandem with recent attempts to incorporate networks into systems theory, or to reconcile Niklas Luhmann's general approach with White. Luhmann has developed an impressive theoretical architecture, where communication proceeds self-referentially, building on previous communication and establishing various kinds of social systems—from face-to-face encounters through formal organizations (companies, universities, administrative units) to large-scale functional subsystems of society like politics, the economy, law, and science (1990, [1984] 1995, [1997] 2012, [1997] 2013). Luhmann thus stresses the self-production (autopoiesis) and production of bounded social entities (systems). He is frequently dismissed as "another systems theorist" in the structural-functionalist tradition of Talcott Parsons. But his approach is less rigid and more open than that of Parsons, and also more iconoclastic with human actors by and large dismissed from his theory. There are many important ideas to draw on, and a number of overlaps with relational sociology (White et al. 2007; Fontdevila, Opazo, and White 2011; Guy 2018). In particular, his focus on communicative process and his notion of meaning (discussed earlier) are of interest here, as well as his relegation of human actors to projection points in the social world and his insistence that pockets of social order have to be accomplished perpetually. However, Luhmann's social world is composed of bounded entities, rather than of relational structures like social ties and networks.

From 2000 onward, systems theorists have probed into how network phenomena fit into Luhmann's theory, to potentially account for them. Veronika Tacke argues that network relations between persons complement the systemic structures of modern society (2000; Bommes and Tacke [2006] 2012). Functional subsystems like the economy, law, and politics render contacts between those spheres attractive and likely, for example, between a politician and a journalist. Boris Holzer follows this lead and locates networks exclusively in modern, functionally differentiated society (2008).

In contrast, authors like Dirk Baecker, Stephan Fuchs, Loet Leydesdorff, and Marco Schmitt argue for a comprehensive theoretical overhaul. Baecker views systems, networks, identities, and culture as "forms of communication," that is, structures of meaning developing in the process of communication, and channeling it in turn (2005, 79f, 226ff). In this vein, Schmitt sees networks and systems as "memories of communication" (2009). Communication leaves various traces that connect to preexisting structures and add to them. Luhmann's systems and White's networks are the most important of these structures.

For Fuchs, in contrast, networks are the master concept (2001a, 191f). He considers systems a special case of networks that increasingly turn inward and establish a boundary of meaning around themselves (2001a, 211ff). Fuchs's theory is mostly about networks of culture, in particular about relations of concepts in scientific theories. He does not really clarify whether, and how, it also applies to *social* networks between individual and collective/corporate actors. Leydesdorff is also mostly concerned with networks in academia (2000, 2010). Building on Luhmann's theory of communication (but also on Habermas and Giddens), he envisions three interrelated kinds of networks: between communicative events (publications), between actors (authors), and between ideas (e.g., operationalized as keywords in publications).

I build on these formulations in different ways. With Baecker and Schmitt, I firmly agree in seeing the process of communication as fundamental, and in placing networks and systems next to each other in this process. In this book, I do not dwell on the relationship between networks and systems, though I envision a certain degree of mutual impact. I agree with Holzer in conceptualizing social relationships as autonomous systems of communication that develop their own structures of expectations (Holzer 2006, 93ff; see chapter 8). Following Fuchs, collective actors can be seen as a special case of networks drawing a symbolic boundary around themselves, making for their reproduction in communicative events (sections 5.9, 8.9; see also Abbott 1995). And I pick up on Leydesdorff in distinguishing *communicative* networks of events, *social* networks between actors, and *cultural* networks of relations between concepts or symbols (section 8.7).

In contrast to most systems theorists, I detail how exactly the process of communication breeds social networks. Also, I examine how social networks interplay with various forms of meaning like relationship frames or social categories. These belong to the realm of culture mostly ignored by Luhmann and most of his followers.

(g) Summary

Clearly, my perspective borrows freely from many of the approaches discussed. Like interactionism, systems theory, and Dépelteau, I look for the processing of social events underlying the formation and change of social structures. My particular conceptualization of these events focuses on the communicative processing of meaning, as in relational sociology around White and in systems theory. The focus on meaning resembles that in the pragmatist and interactionist approaches, and in relational sociology. All of the approaches sketched here contain some important impulses to build on. Since my main goal is to lay out my own theoretical perspective, I pick up selectively on alternative approaches rather than discuss them at length. From time to time, I highlight important differences both to other authors of relational sociology, and to alternative approaches in the social sciences.

1.5. Outline of the book

The book starts with the interplay of meaning and networks. Chapter 2 offers the basic framework of social networks as dually consisting in (a) *observable regularities in communication*, and (b) of a *"meaning structure"* (chapter 2). The meaning structure features the identities of the actors involved and "relational expectations" about their behavior toward each other. Such expectations (and the identities involved) arise from the process of communication and channel it effectively, making for the communicative regularities. These ideas are applied to the case of *interpersonal relationships* and their negotiation by recourse to cultural frames like love and friendship. This connects to Zelizer's notion of "relational work."

The third chapter hones in on *groups* and *social boundaries*. Groups are special cases of cohesive network structure. They develop their distinct symbolic repertoires (group cultures). For their reproduction, they depend on regular meeting places (foci of activity) and on the construction of a group boundary, marking the group as different from other groups and from the outside world. These boundaries have an imprint on the relational expectations between individuals, prescribing collaboration and solidarity within the group, conflict and competition to other groups, and indifference or rebellion to larger social and political structures. In turn, groups depend on interpersonal relationships following these symbolic demarcations.

Chapter 4 extends these ideas to *ethnic categories* and *cultural differences* of migrant groups (section 4). Ethnic categories (including "race") structure the interaction within and across the category, fostering strong personal ties within the group and more superficial routinized interaction with members of other categories. This reinforces distinct ethnic cultures, which, in turn, make for the classification according to ethnicity, and for expectations about the behavior of members of ethnic groups. But this circular reproduction of ethnic differences also depends on other factors, such as the meeting of members of ethnic groups in neighborhoods, schools, universities, clubs, or at the workplace. Also, ethnic categories cannot only map and symbolically separate cohesive groups. They can also serve for the exclusion of dispersed outsiders from a cohesive in-group, as in established–outsider configurations (Elias and Scotson [1965] 1990).

Chapter 5 turns to roles and institutions as a third instance of the symbolic patterning of social networks. I discuss various sociological approaches (role theory, German philosophical anthropology, Berger and Luckmann's sociology of knowledge, neo-institutionalism, and relational sociology) with regard to their conceptualization of roles and institutions, and to their systematic connections with social relationships and networks. *Institutions* are considered as cultural models for the organization of interaction. Relationship frames, social categories (e.g., gender), and formal organization are "relational institutions" that pertain to the patterning of social relationships in networks. The *role* concept stands the particular patterning of network ties by recognizable positions (following blockmodel analysis). These positions and the relations between them partly follow institutional cultural models (like the family), but they can also emerge endogenously from the interaction in the network.

The sixth chapter builds on chapters 2 and 5 to discuss *love* as a relationship frame, and *gender* as a social category making for a cultural imprint of personal relationships. I see both as intertwined, with gender as the social categories tied to the traditional ideal of heterosexual love. Also, love and gender are subject to (changing and context-dependent) cultural construction and to interpersonal negotiation. In tandem, they structure interpersonal relationships by structural equivalence, making for heterosexual love ties and gender-homophilous friendships, as well as the incorporation of couples into their respective family networks.

Chapters 7 and 8 move from a relatively static view of the connection between networks and meaning to their dynamics in the *process of communication*. Chapter 7 discusses various alternative conceptualizations of the *events in social networks* with regard to their assumptions, their methodological implications, and their compatibility with the relational sociological perspective: behavior (in the sense of Homans), action (Weber), practices (Bourdieu), exchange (Blau, Emerson), interaction (Mead, Blumer, Goffman), transactions (Emirbayer, Tilly, Dépelteau), and switchings (White). From a comparison of these concepts, I argue for focusing on processes between actors, for avoiding recourse to unobservable subjective processes, and for conceptualizing and examining the processing of meaning (in these events).

As laid out in chapter 8, these requirements lead to Niklas Luhmann's *theory of communication* as the (almost) perfect concept for the processes in networks. I argue that communicative events are routinely attributed to actors, leading to expectations about their identities and their behavior toward each other. Communication thus generates "relational definitions of the situation." Future communication builds on these relational expectations and continuously confirms or modifies them. These ideas are applied to the problems of intercultural communication and to the construction of corporate or collective actors (and of the relations between them) in communication.

Chapter 9 summarizes the core arguments of the book and reflects on future opportunities and challenges. One area of extension lies in methods that study social networks and patterns of meaning in the process of communication. Apart from the qualitative study of communication, this leads to recent developments in computational social science where social ties and cultural formations are reconstructed from process-generated data. Other areas of extension include the sociology of formal organizations with the interplay of informal relationships and formalized roles and institutions. Also, the theory of fields already accords a prominent role to relations and networks. How does this square with the perspective at hand?

I have tried to make the chapters readable as stand-alone pieces that deal with a single part of the theory, without the readers necessarily having worked their way through, or accepted, other parts. Also, social networks and meaning can be conceptually intertwined, as in chapters 2–6, without connecting them to the notion of communication. These chapters can be read with other concepts for social processes in mind: social action, interaction, transactions, and exchange. The same does not hold for the communication theoretical perspective laid out in chapters 7 and 8. Communication and networks do not really link without reference to meaning. To blend all of this in a coherent and logically integrated perspective, I reiterate some basic points a number of times. Hopefully, encountering some core ideas at various places in the book will not tire readers excessively. I also added a number of cross references to other chapters and sections in this book, in an effort to connect them to each other.

2

Networks, Relationships, and Meaning

As I argue in chapter 1, most network research pays little attention to the expectations, symbols, schemata, and cultural practices inscribed in interpersonal structures: the meaning connected to networks and relationships. This chapter offers a theoretical sketch of the interplay of structure and meaning in networks and of its methodological implications.[1] This lays out the foundations for the coming chapters, going deeper into particular forms of meaning and their interweaving with sociocultural constellations: groups (Chapter 3), ethnic categories (Chapter 4), role patterns (Chapter 5), and love and gender (Chapter 6). My approach draws heavily on the relational sociology of Harrison White and others (White 1992, 2008; Pachucki and Breiger 2010; Mische 2011; Fuhse 2015b; McLean 2017). In addition, I incorporate insights from diverse research traditions such as systems theory, social psychological research on relationships, symbolic interactionism, and social anthropology.

The first section argues on a general plane that social structure is substantively composed of symbolic constructs such as expectations, identities, and categories (section 2.1). Therefore, network research has to deal with the interplay of structure and meaning. The second section gives a brief excursus on the notions of culture and meaning (section 2.2). A historical overview follows of how culture was incorporated into the different strands of network research (section 2.3). The fourth section argues that social relationships as the constituent units of networks consist of relational expectations. These are symbolic in nature and deeply interwoven with cultural forms (section 2.4).

Sections 2.5–2.8 examine various ways in which dyadic relationships are coupled to each other to form a network. First, multiple ties frequently become relevant in one series of communicative events (section 2.5). Second,

---

[1] This chapter is a substantially rewritten and extended version of my article "The Meaning Structure of Social Networks" published in *Sociological Theory* 27/1 (2009). I would like to thank *Sociological Theory* and John Wiley and Sons for the permission to reuse the article.

Social Networks of Meaning and Communication. Jan Fuhse, Oxford University Press. © Oxford University Press 2022.
DOI: 10.1093/oso/9780190275433.003.0002

identities are constructed as involved in multiple relationships (section 2.6). Third, social relationships (and the identities connected to them) are frequently negotiated and defined by recourse to culturally available "relationship frames" (section 2.7). These offer models for relationships, but also for their embedding in network patterns. Finally, network constellations correspond to cultural formations (section 2.8): symbolic forms tend to be shared in densely connected network clusters, but also by structurally equivalent positions in a network.

2.1. Levels of network research

This section sketches the overall framework for the following arguments. Research on social networks deals with at least five interrelated levels (Figure 2.1):

(1) The first level is the *analytical picture* of the network as derived from empirical observation with the respective research methods. On this level, researchers often deal with 1s and 0s, for existing and nonexistent dyadic relationships between actors. The 1s and 0s are, of course, rough abstractions of a complex social reality: a relationship is observed to exist or not to exist—depending on operational definitions of relationships. What a relationship actually is, or whether

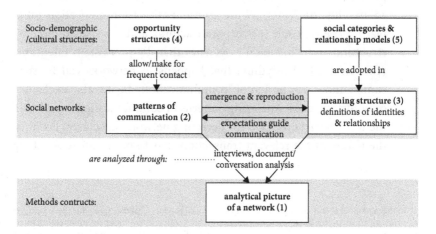

Figure 2.1 Integrated model of network research.

the relationships observed differ fundamentally, has to be bracketed for analytical purposes. For example, the friendships among employees in a company probably all involve different activities and carry a different meaning for the people involved (Krackhardt 1999). But they have to be treated alike in formal network analysis. The same holds for non-ties. Both the 1s and the 0s in a network matrix have to stand for roughly similar phenomena, respectively.

This highly abstracted view on social structures builds on networks as a construct of network analytical research methods, and it allows for a wide array of very fruitful research, examining the overall structure of relationships, or individual variance in the composition of ego-centered networks. It is useful, though, to look behind the abstractions and assumptions of this analytical picture—at what the picture is designed to depict. Doing so can give us answers to structural puzzles as to why we find different network structures in different contexts (or by different types of tie), or why seemingly cohesive network structures disintegrate (Yeung 2005; Gondal and McLean 2013). The analytical construct of a social network is usually based on observations of two other levels of social networks: observable patterns of communication and expectations between actors.

(2) Patterns of *communication* are analyzed, for example, in trade networks (Erikson and Bearman 2006), in the sending of emails in a university (Kossinets and Watts 2009), or through recording conversation in meetings (Gibson 2005).[2] Transaction or communication processes have a distinctly supra-personal quality (Luhmann 2002, 155ff; Tilly 2002, 48f; White 1995a, 1037). Communication takes place *between* people (or corporate actors), and it is shaped more by structural and situational exigencies than by preexistent intentions (Gibson 2000). Everything that happens in networks can be subsumed under this notion of "communication," be it verbal exchange, or nonverbal cues. Analyzing patterns of communication is probably the most reliable method for detecting network structures. For this, the nature of the relevant transactions has to be specified. Are they

---

[2] In the version of this chapter published in 2009, I used the term "transactions" for these patterns of communication. In this book, these considerations are framed in the wider communication theoretical perspective (see chapter 8). The precise conceptualization of the communicative processes in social networks is irrelevant for the arguments here.

cooperative or conflicting; is exchange mutual, equal, even, unbalanced, or unidirectional?

(3) While communication is *what* happens in networks, the third level is *why* particular communication occurs between actors. This is the level of interpersonal (or inter-actor) expectations that evolve in and guide communication, or of the *meaning structure* of social networks. In a sense, all social structures consist of expectations—of a pay cheque in exchange for work, of traffic rules, of legal sanctions to criminal activity, etc. Social networks and relationships feature specific *relational* expectations. These pertain to how one actor will and should behave toward *particular* other actors, rather than to just anybody, as in traffic rules or the prospect of legal sanctions. An email from A to B, for example, springs either from formal arrangements such as the professional rules of professor and student. Or it might come, say, between two students, out of an encounter in a seminar or at a party, and then lead to follow-up email communication between A and B. Communication is guided by expectations, be they mostly formal (professor–student) or mostly informal (student–student).

The same holds in any other network ties. Leaving formal arrangements aside, relatively stable patterns of communication emerge through establishing a structure of expectations concerning the behavior of particular actors toward each other in the network. These I term "relational expectations"—since they do not reside in formal organizations (like the relation between professor and student) or in larger social fields (like the tie between buyer and seller in an economic market), but only in the tie between two actors. By virtue of repeated communication and relational expectations, the tie between A and B becomes a relation*ship*.

Relational expectations form part of the "meaning structure" of a network since they are inherently based on meaning. They make sense of previous encounters by establishing definitions of "what is going on" (Goffman [1974] 1986). And they lead to further communication, prescribing conflict, cooperation, loyalty, or obedience. This meaning structure of a network consists of definitions of relationships as a particular type of tie (love, friendship, competition, conflict, etc.) and of the construction of identities of the actors involved (as related to other actors; Wood 1982; Somers 1994; Yeung 2005). Often, these expectations result from complex negotiations of relationships and of identities (McLean 1998, 2007).

Definition 2.1. The meaning structure of a social network consists of the forms of meaning (identities, stories, expectations) attached to actors in a network, and to the ties between them.

The meaning structure of a network is a fuzzy social reality that has been little theorized. As I argue in the next section, it is subject to two important distinctions: First, we can locate meaning either *subjectively* on the level of individual perception and expectations, following Max Weber and Alfred Schütz. Or we can examine meaning *in communication* as interpersonally established expectations and symbols, in the sense of Niklas Luhmann. Later in this book, I argue for the theoretical and methodological primacy of meaning that is communicated, rather than only held subjectively (see section 8.3; chapter 9). Suffice it here to say that meaning is processed both subjectively and in communication, and the two interplay in complex ways (DiMaggio 1997). Every network context produces its own web of intersubjectively shared expectations. But actors also carry these expectations (and other symbolic forms) "in their heads" from context to context.

The second distinction concerns the locale of meaning in the social world: social relationships and identities carry a particular meaning for the actors involved and in the communication in their *immediate context*. This is what I subsume under the term "meaning structure" of the network—the relational expectations concerning the behavior of particular actors toward each other. But this meaning structure almost always draws on the repertoire of symbolic forms as available in the *wider cultural context*. In this sense, relationship frames like "friendship," social categories like "gender," and roles like "patron" and "client" come with a particular meaning that is more widely distributed and used than the immediate micro-social context. As I argue in chapter 5, they are institutionalized models for the formation of social relationships. This distinction translates into that between cultural structures like social categories and relationship models (level 5; see further discussion) and the meaning structure of the network itself (level 3). While the meaning structure of the network may feature symbolic forms from the wider cultural context, the two should be held separate conceptually and analytically.

How can the meaning structure of social networks be analyzed? *Subjective* conceptions and expectations are most often the basis of network research. Network generators, like those developed by Claude Fischer (1982a, 284ff) and Ronald Burt (1984), draw on individual accounts of network ties. When analyzed on this level, the network is equated with the aggregate of its subjective perceptions by the actors interviewed. These subjective accounts in interviews often show remarkable differences in the understanding of categories (Bearman and Parigi 2004; Fischer 1982b) and are, of course, influenced by the interview situation (Sudman, Bradburn, and Schwarz 1996, 55ff). This leads, among other problems, to remarkable levels of asymmetry in supposedly reciprocal ties, for example, in friendship nominations. The *communicative* side of the meaning structure is more difficult to analyze—it cannot be tapped in interviews. In order to grasp intersubjectively constructed relationship definitions and relational identities, one has to analyze the content of transactions. This can be done through the analysis of documents or of speech acts (Gibson 2005; McLean 1998; Mützel 2002; McFarland 2004; Fuhse et al. 2020, 10ff; see section 8.10).

The distinction offered between patterns of communication and the meaning structure of social networks mirrors that of network ties as measured as relational "states" (e.g., friendship, liking) and on the basis of "events" (e.g., sending email, giving advice) by Stephen Borgatti and Daniel Halgin (2011, 1170) and that between "role relations" (e.g., friendship) and "behavioral interactions" (e.g., discussing important matters) by James Kitts (2014). The two sides (relational states / role relations as part of the meaning structure and events / behavioral interactions / communication patterns) are closely tied to each other: relational states develop from events and structure them. Both constitute valid measurements of the complex underlying reality of network ties, focusing on different aspects of it.

If meaning structure and patterns of communication are so intricately linked, why disconnect them conceptually? Of course, any two features of the social world never perfectly match, for example, when communication does not follow expectations, or when these gradually build up over a sequence of communication. This is why Kitts admonishes keeping the two kinds of measurement and conceptualizations apart, to avoid conflation and confusion (2014). Meaning structure and communication patterns are also influenced by different factors. I focus here on opportunity structures, social categories, and relationship models. These order networks by constraining and enabling particular communication patterns or meaning structures.

(4) *Opportunity structures* for contact include the place of residence and other foci of activity like bars, sports clubs, schools, or workplaces (Feld

1981). They are sometimes used as proxies for measuring actual ties but really influence them as part of wider social structures (Kitts 2014, 278ff). Foci of activity have an impact on the communication taking place, because people meet more often when living in the same neighborhood, attending the same school, or frequenting similar bars or sports clubs. This can lead to the formation of new ties, or to the deepening of existent relationships (Mollenhorst, Volker, and Flap 2014).

Place of residence and other foci of activity are characterized by particular compositions of people meeting there in terms of age, affluence, ethnicity, gender, and lifestyle. They frequently foster networks that are homogeneous with regard to these attributes. While more structural than cultural, opportunity structures are often not devoid of meaning. For example, people may move to a particular neighborhood because they expect to meet certain people there. Similarly, bars or sports are connected to specific lifestyles and thus bear a distinct meaning for the people who attend them, and for the interactions taking place.

(5) The second group of factors comprises *relationship models* (like love, kinship, and friendship) and *social categories* (gender, age, ethnic descent). Both belong to the realm of *culture* in the sense of forms of meaning that are shared in a particular context (see section 2.2). In the terminology from chapter 5, they are "relational institutions"—models for how to relate. Relationships models and social categories make for a structuring of ties and networks according to culturally prescribed expectations (Rytina and Morgan 1982; Yeung 2005). Of course, people do not have to adhere to these models. But they make relationship life considerably easier by supplying prepackaged expectations that communication can draw on to reduce uncertainty. If the expectations attached to relationships models and social categories are adopted, they become part of the meaning structure of social networks, indirectly affecting communication patterns.

Both opportunity structures and the relational institutions of social categories and relationship models form part of the wider sociocultural configuration of society, including socioeconomic inequalities, housing patterns, cultural lifestyles, and relationship ideals, as well as the symbolic status orders of gender, class, caste, and race/ethnicity. Actual networks are patterned by the combination of these factors. Opportunity structures enable (or prevent) contact in the first place. Categories and relationship models, in contrast, prescribe the interaction taking place, and order contacts into patterns of structural equivalence (Lorrain and White 1971).

A heterosexual singles dance venue, for example, provides opportunities of contact (level 4) as a focus of activity (Beattie et al. 2005). But only the application of the gender category (level 5) will usually make sure that male-female dyads emerge on the dance floor. Similar considerations hold for the caste system. Indian villages offer abundant opportunities for contact. But the actual contacts, and the relationships that form, still follow by and large the traditional norms for interaction between members of different castes (Sharma 1999; Arora and Sanditov 2015).

As sketched in figure 2.1, the causal starting points are the sociodemographic and cultural structures. Opportunity structures (4) like the neighborhood, the workplace, school, clubs, and informal meeting points make for a structuring of the patterns of communication (2). Social categories and relationship models (5), in contrast, are adopted as part of the meaning structure (3). They influence the definition of identities and of relationships among the actors involved. Communication patterns and meaning structure, together, constitute the network as a *real social structure* (as opposed to its *analytical representation*; 1). They are intricately tied to each other: the definitions of identities and relationships emerge, reproduce, and change over the course of communication. The expectations embodied in these definitions in turn guide future communication.

> Assumption 2.1. In social networks, their meaning structure and the observable patterns of communication are tightly interconnected.

Assumption 2.1 formulates the basic principle for the chapters 2–6 of this book. Communication patterns and structures of meaning are interwoven and therefore connected to each other. It is possible to study this interplay empirically, making this assumption lean toward a very general proposition. However, the proposition has to be specified to be tested. And often enough, we will not be able to relate these two levels of networks empirically. Therefore, I formulate this as an assumption from which to derive testable conjectures.

Of course, there is feedback from networks to sociodemographics and cultural patterns. Social categories and relationship models are reproduced and even modified in the interplay of communication and meaning structure.

For example, the nature of the gender category might change as a result of the ongoing negotiations in everyday life (Ridgeway and Smith-Lovin 1999; see chapter 6). Or members of a subculture increase the probability of meeting each other (opportunity structure) by moving into specific neighborhoods.

Communication and meaning—as the "real" structure of a network—are mapped in the analytical picture of a network (1). This "picture" can consist of network matrices or of their graphic representation. The means of acquiring these network data are interviews and the analysis of documents or conversation. Network analyses draws (statistical) connections between sociodemographics and cultural patterns (4 and 5), on the one hand, and this analytical picture of the network (1) on the other hand. A *theory* of social networks, however, has to deal with the mechanisms linking the various levels of network research in more detail.

2.2. Excursus on culture and meaning

Before going on, I have to clarify the notions of culture and meaning. Both concepts, though central to sociology and anthropology, are notoriously hard to define. I offer a short guide to how the two terms are used here, and how they differ from each other.

(a) Meaning

In spite of the prominence of the concept of culture, I regard "meaning" as the wider term. Following the traditions of phenomenology and interpretive sociology, humans attach meaning to objects and to symbols, effectively acting on these meanings (rather than on the objects and signs themselves). According to Max Weber, all action is driven by the subjective meaning individuals hold of their environment and of alternative courses of action ([1922] 1978, 22ff). Alfred Schütz insists that an actor orients herself to various layers of meaning—of the act itself, of other actors, of the presumed meaningful orientation of other actors, and of what meaning others will attach to the actor's behavior—effectively anticipating their expectations and reactions ([1932] 1967). In effect, then, we are dealing with twin versions of reality: the objective world and our ideas of it. For example, car drivers do not stop at a crossing because a pole with a red light somehow blocks their way, but because of the meaning they connect to it. They see it as a red traffic light, part of a wider system of traffic rules that specify what they are supposed to do.

Laws, the family, economic enterprises, churches and sects, ethnic groups, political parties, and states are all crystallized structures of meaning that a large number of people similarly hold and act upon. All of these are not objectively given but exist and persist only as constructions of meaning. Individuals have to orient their actions toward these constructions as others are expected to do the same, and this gives them a quasi-objective quality. They become "objective meaning," as Schütz formulates. Human actors here act upon the "expectations" they hold of the behavior of others (Weber [1913] 1981, 159ff). In a sense, all social structures consist of expectations (Parsons et al. [1951] 1959, 19f; Luhmann [1984] 1995, 96f). These include both cognitive expectations as to what will happen, and normative prescriptions of what should happen, as well as expectations about the expectations of others (Galtung 1959; see section 7.3). How will and should others act, and how will they react to my different alternative courses of action?

These considerations connect well with arguments from American pragmatism and symbolic interactionism: Individuals subjectively process symbols and negotiate their meaning in interaction (Mead [1934] 1967; Blumer 1969). From this process, a "definition of the situation" arises for subsequent interaction to start from (Thomas [1927] 1966). One of the examples that William Thomas ([1927] 1966, 166) and Robert Merton (1958, 477ff) offer for the "definition of the situation" (one also explored by Weber) is the ethnic categorization that divides people into groups on the basis of believed common descent. Clearly, meaning is not confined to individual minds, but communicated and shared to become consequential.

I build here on an important twist added by Niklas Luhmann: he regards meaning as processed dually, both in psychic systems (in people's minds) and in communication ([1984] 1995, 59ff; see section 8.3). Luhmann argues that both are intertwined, but that sociology should focus on meaning *in communication*. While people have their own thoughts, only communicated meaning becomes consequential in the form of organizational decisions, opinions expressed, mass-mediated news, scientific statements, political claims, etc. Traffic lights, then, do not make car drivers stop when red because they independently arrive at the subjective idea that they should do so—but because the meaning of red traffic lights is defined in the web of communication, and because ignoring them cannot only lead to accidents but also to follow-up communication, including legal sanctions.

This focus on communication radically breaks with the action-theoretical tradition of Weber and Schütz (where the meaning of an act is subjectively intended by the individual actor). But it connects with the dual processing of symbols in thoughts and in interaction, as devised by Mead and Blumer. Most symbolic interactionism, though, focuses on interpreting subjective meaning, rather than meaning-in-interaction. And Luhmann's position mirrors the methodological stance of conversation analysis to study the meaning of linguistic forms and communicative events (see section 8.2). While locating meaning outside of the individual sounds outlandish at first, it is not that different from the positions of Michel Foucault (meaning in discourse), Erving Goffman (dramaturgical meaning), Mary Douglas (meaning in institutions), Pierre Bourdieu (meaning in practices), and neo-institutionalism (meaning in rules of fields; Wuthnow 1989). Also, the traditions of semiotics and more generally, linguistics, consider meaning in signs, rather than only in people's minds (de Saussure [1916] 2013; Morris 1971). Meaning is located and analyzed in communication (verbal or in texts) without inferring the subjective dispositions and interpretations of actors, quite in line with Luhmann's notion.

What phenomena exactly are covered by the concept of meaning? First, we rarely find meaning in isolated form. Rather, the meaning of something— a symbol, a concept, an action, an event, a sentence—always relates to the meaning of other things (other symbols, concepts, actions, events, sentences). Red traffic lights carry a different meaning than green traffic lights or a stop sign. To take another example, we can interpret the meaning of makeup by comparing it with similar and dissimilar behavior. Wearing makeup has a different meaning than wearing none. But it is also related to the event in question, related to social categories (gender, age, sexuality), and compared with other kinds of makeup worn for the same occasion. Such observations are not only academic but make for an important part of the social world, especially when they are explicated and negotiated in communication: "You could at least have made an effort for the wedding!" "Is Steve really wearing eyeliner?" Similarly, the meaning of a word consists in the relations to other words (de Saussure [1916] 2013). That is why we define concepts with other words—because a word in itself does not carry meaning. In this sense, Clifford Geertz speaks of human beings as "suspended in webs of significance" (1973, 5). Meaning is organized in relations connecting symbols, practices, social contexts, roles, and expectations, to each other (Fuchs 2001a; DiMaggio 2011).

However, regarding meaning as a network or "web" does little to specify a designated area of social phenomena. The various approaches to meaning cover an extremely wide array of aspects of the social world. Meaning comprises symbols, concepts, intentions, perceptions, norms, narratives, and values in thoughts, in talk, and in documents. Social structures carry meaning, just as symbols and sentences do. Following Weber and Schütz, the term captures the way in which human actors make sense of the world and orient to it (in terms of concepts, symbols that relate to each other as well as to the world around them). Luhmann tells us that communication similarly represents the world, making up its own version of the world as a meaningful construct. In both approaches, pretty much everything in the social world qualifies as meaning—whether in the meaningful orientations of individuals or in the communicative construction of social structures.

Overall, the notion of "meaning" signifies an *interpretive stance* in the phenomenological tradition. Social scientists are supposed to investigate what people think of the social (and the natural) world. Rather than look for systematic connections between social facts ("explanation"), sociologists inquiring into meaning pursue an "understanding" of reasoning and rationales for action, but also of the social structures constructed in joint, coordinated action (Weber [1913] 1981). Meaning covers what we see (or strive to see) when interpreting signs, texts, or behavior. Meaning is not a thing out there, but the focus and result of a research perspective.

Definition 2.2. Meaning is the web of symbolic associations in thoughts and in communication. We observe meaning by adopting an interpretive stance.

The basic ideas of Weber, Schütz, Thomas, and Merton are by now rather commonsensical in sociological theory (even Luhmann's focus on communication has devoted followers). We cannot regard social structures from relationships and families through formal organizations and collective identities (like sects or social movements) to states and economic markets as objective structures. They are symbolically constructed and vary in the meaning we attach to them. They do not exist over and above symbols, values, expectations, institutions, and other cultural stuff—rather, their social reality lies in structures of meaning we devise (subjectively and in communication), and in their effectively guiding the behavior of the actors involved. The same should hold for social ties and networks—they rely on "definitions of the situation"; they consist of expectations and of the meaning attributed to them.

These are the teachings of sociological theory, but network researchers rarely incorporate them in their analyses and writings. This may come from the additional hurdles that interpretation brings into structural analyses. If we can study social networks as acultural structures—why bother with meaning? The structuralist stance teams up with the widespread theory abstention in network research, as sketched in chapter 1. After all, asking what social ties and networks are, or what they are made of, only complicates matters and does not necessarily make publishing in *Science* or *Social Networks* more likely. I make the opposite case: that disregarding meaning in relationships and networks is overly facile and leads to faulty interpretations and research designs. Incorporating meaning in theory and in empirical research does make both more complicated by adding another layer of complexity. But it leads to more interesting observations and to better results.

(b)   Culture

The concept of culture is slightly easier to define than meaning, though it suffers from similar breadth. Disregarding the narrow equation with "high culture" (art, classical music, literature, philosophy), the wide notion of culture encompasses all sorts of forms of meaning—norms, values, knowledge—as well as language and other symbols, but also practices (custom, mores, etc.). What, then, is the difference from the term "meaning"? Culture is always *shared* within a certain context, and then we call it the "culture of this context." If one person holds a specific belief, we would not call it culture. But if a number of them do, we can speak of this belief as cultural.

In this sense, Harrison White views culture as a "continuously interacting population of interpretive forms within some social formation" (1992, 289). Margaret Archer defines culture as "the contents of libraries" ([1988] 1996, xvii). This might be too lax: All sorts of ideas, languages never spoken, and long-forgotten words linger in libraries. Much street jargon, many musical themes and cooking recipes, in contrast, have not made it into libraries. Archer seems not overly concerned with the social appropriation of cultural forms but focuses on the inner logical consistency of "cultural systems" ([1988] 1996, 104ff). I am not interested in culture as a metaphysical level floating above the social world, but in the connection between social and cultural patterns. Following the anthropological tradition, I conceptualize culture as not only available, but also as *widely used*.

Definition 2.3. Culture denotes the repertoire of forms of meaning that are available and widely used in a specific context.

In any case, the forms of meaning in a "culture" have to be institutionalized, that is, become independent of their particular application in communication and acquire an abstract meaning of their own. For example, a handshake means something (a friendly, but relatively formal greeting, among other things) across a wide variety of situations and actors, even if nobody shakes hands for a while. I consider this institutionalization of cultural forms in more detail in chapter 5. Suffice it to say here that cultural forms are known and relatively invariant in their particular context.

The social habitats of culture range from small to large. Dyadic relationships already develop their own jargon and their distinct ways of communicating about the social world—a "relationship culture" (Wood 1982). Similarly, we speak of the culture of an organization, of delimited social groups like class, ethnic groups, or castes, or of the population in a subnational, national, or supranational territory (e.g., New York City, Germany, Latin America). Importantly, speaking of forms of meaning as cultural, or as part of the culture of some social context, always compares this context (with regard to the forms of meaning available and widely used) to other contexts (Luhmann 1995a, 31ff).

If we take social networks as the contexts of culture, the concept becomes synonymous with White's notion of "domain" (1995a, 1038f; Mische and White 1998, 702ff). The domain comprises all sorts of symbols, narratives, and linguistic forms that emerge in a densely knit network context, characterizing it in contrast to other network contexts (see section 2.8). Certainly, forms of meaning are diffused and reproduced in the communication in networks. But culture is also transmitted through other means, like the mass media, institutions of learning and socialization, and economic markets (Hannerz 1992; Stevenson 2002; Fuhse 2018b).

Note that the definition presented in this chapter does not really side in the debate between *determinist and agentic notions of culture*. Social contexts are only relatively homogeneous with regard to the forms of meaning. With more fine-tuning, it is always possible to find alternative symbols, practices, and values to choose from, in the sense of Ann Swidler's "culture as tool-kit" metaphor (1986). However, it is impossible to draw on cultural material *not available* in a given context. Relatively isolated and close-knit network contexts will therefore display more homogeneity, as in Geertz's anthropological studies (1973). Contemporary society, in contrast, almost always provides access to diverse forms of meaning, from crosscutting social circles, from the mass media, or from other channels of communication and cultural diffusion. The extent to which culture determines action and communication, or that cultural forms can be creatively combined, or even subverted, depends on the configuration of sociocultural context (DiMaggio 1997).

In relation to the framework from section 2.1, culture includes the social categories and relationship models from level 5. These are drawn upon for the negotiation of social relationships and identities, for example, in terms of the cultural model of "love" and of the role "husband." This is the level of meaning assigned to actors and to the ties between them (level 3). However, I do not speak of the meaning of a network here, but of the network as a "meaning structure." This shift stresses the bundled and relatively stable nature of these meanings. Relationships and networks are structures that usually do not change from one moment to the next just because someone attaches a different meaning to them. Rather, they resemble formal organizations, sects, families, and other social structures as crystallized and persistent patterns of meaning that effectively constrain individual behavior—because we can expect others to behave according to these structures of expectations.

2.3. Incorporating culture

With these conceptual clarifications, we can examine how culture and meaning were incorporated, or ignored, in research on social networks (McLean 2017). The original network research from the 1950s well into the 1980s was little concerned with cultural forms. British anthropologists J.A. Barnes and Elizabeth Bott invented the network concept to look behind apparent self-descriptions of societies as structured by categories. The original 1954 article of Barnes describes a Norwegian island parish as much less characterized by class divisions than by a relatively homogeneous mesh of personal relationships. In a similar vein, Bott shows that the gender division of labor in British families depended on the structure of their wider social networks ([1957] 1971). On the other hand, Philip and Iona Mayer point out that different lifestyles of urban migrants in South Africa were rooted in a different organization of social ties (1961). Around 1970, J. Clyde Mitchell called for an inclusion of aspects of meaning in social anthropology, especially with regard to norms and the subjective meaning of social relationships (1969, 20ff; 1973, 26ff). In the 1990s, Ulf Hannerz (1992) and Thomas Schweizer (1996) wedded their social anthropological works with a systematic concern for cultural aspects.

A similar process of incorporation of the level of meaning into network research took place in sociological network research. Early sociometric and network scholars were mainly concerned with methodical refinement. The crucial "breakthrough" was accomplished at Harvard in the 1970s, when Harrison White and his colleagues developed the blockmodel technique (Scott 2000, 33ff; Breiger 2004, 510ff; Freeman 2004, 121ff). Building on structuralist studies of kinship, blockmodeling aims at reconstructing categories of actors with similar connections (to similar other actors; Boyd 1969; Lorrain and White 1971; Lorrain 1975, 107ff). Two aspects of the method are of interest here:

- Blockmodeling starts from the idea that network patterns follow underlying categories of actors and make for the systematic relations between these categories. Rather than investigate the use of such categories, blockmodels inductively generate them from formal network data (Boorman and White 1976; White, Boorman, and Breiger 1976; see chapter 5). Like Barnes, they start from an "anti-categorical imperative" (Emirbayer and Goodwin 1994, 1414): Culturally available and socially diffused categories are mistrusted. The "real" categories and the patterns between them are inferred from network structure. While blockmodeling tries to map the categorical order of social networks, it does so by *ignoring preexisting categories and self-descriptions.*
- The blockmodel technique builds on the distinction of *types of tie*: "fathers" are typically connected to "mothers" through the tie "marriage," and to their children through "parenthood." In their analysis of the network of monks in a monastery, White and his coauthors distinguish between eight types of tie between the monks, from "like" and "esteem" to "antagonism," "disesteem," and "blame" (1976, 749ff). Thus, blockmodeling derives categories of actors from network data based on the symbolic differentiation of different types of tie. However, as the authors note, "The cultural and social-psychological meanings of actual ties are largely bypassed" (1976, 734)

Culture and meaning thus linger in blockmodeling (in categories and in types of tie; Brint 1992; DiMaggio 1992, 120f). But they are ignored in actual research practice. White himself turned to the theoretical reflection of network research around 1990, out of a frustration with the purely structural understanding of networks and ties (Mische 2011, 82). In his 1992 theoretical opus magnum, *Identity and Control*, he terms networks "phenomenological realities" and "networks of meaning" (1992, 65, 67), composed of "stories" and "identities." His particular interest lies with the role of language in organizing social networks (1992, 133ff; 1995b, 714ff; 1997, 57ff). In contrast to his earlier work on blockmodeling, White's *Identity and Control* promotes culture and meaning to levels of research in their own right, rather than merely underlying the network matrices of blockmodels. However, he considers cultural forms or patterns only in their inscription in concrete social structures, rather than as abstracted from particular interaction and relationships and institutionalized.[3]

A number of authors around White have followed his general lead in the "(New York school of) relational sociology" (Pachucki and Breiger 2010; Mische 2011; Fuhse 2015b; McLean 2017). Their approach focuses on the theoretical reflection of social networks, and on their application to sociocultural phenomena Mustafa Emirbayer and Jeff Goodwin argue for increased attention to the symbolic construction of categories and to the subjective meaning of actors (1994). Influenced by White and Emirbayer, Charles Tilly examined the narrative construction of political identities and social boundaries in networks (1998, 2002, 2005a). Several authors investigate the cultural underpinnings of network patterns (DiMaggio 1992; Ikegami 2005; Gondal and McLean 2013) while others examine the negotiation of identities and relations in networks (Somers 1994; McLean 2007).

Some relational sociologists have pursued this line to analyzing networks of categories or of relationship frames, instead of actors, with formal network analyses (Mohr 1994; Martin 2000; Yeung 2005; DiMaggio 2011). According to John Mohr, the aim of this research is to identify patterns of culture (1998). It is concerned with *cultural networks* between linguistic terms, no longer directly with social networks of individuals or collective actors. This now

---

[3] Mustafa Emirbayer and Jeff Goodwin criticize White in this regard (1994, 1439ff), but without detailing why cultural forms could detach from social networks. Margaret Archer argues that the cultural system is integrated by "invariant logical principles" ([1988] 1996, 104). To avoid assuming a metaphysical level of reality, I argue for the relative independence of cultural forms from social networks due to their different channels of transmission and negotiation (Hannerz 1992).

makes for a thriving research area, in spite of a widespread lack of theoretical reflection (Light 2014; Rule, Cointet, and Bearman 2015; Evans and Aceves 2016; Fuhse et al. 2020). What does it mean for two concepts to be related, or for a social category to occupy a role in a cultural network? Also, whether cultural and social networks show similar structural properties remains to be seen.

Much of North American network research, however, has continued the structuralist path of ignoring meaning. For example, Peter Bearman follows the "anti-categorical imperative" in arguing that social networks need not be structured by socially known or diffused categories (1997). His analysis of the network of bride exchange among island Aborigines yields the picture of a "generalized exchange" between blocks that did not correspond to any culturally denominated categories. Similarly, the prominent works of Ronald Burt (1992), Nicholas Christakis and James Fowler (2009), and Lee Rainie and Barry Wellman (2012); the physical network research by Albert-Laszlo Barabási, Mark Newman, and Duncan Watts; and most articles published in the journal *Social Networks* do not consider meaning in networks, or they infer "the meaning" of relational patterns simply from the network structure, without addressing the level of meaning in empirical research. Recently, Mark Granovetter reflected that "it seemed very important [for network analysis] to abstract away from normative frameworks" and ideas, but that this led to "the failure [. . .] to sufficiently appreciate and explicitly analyze the important role of larger cultural and political forces" (2007, 4).

Thus, we find within American network analysis a bifurcation between a more structuralist understanding of networks, and one that treats culture as an important aspect of social structure (Breiger 2004, 518ff). The second position postulates that social networks are not only inextricably tied to the level of meaning, but inherently cultural: "social networks are seen not merely as locations for, or conduits of, cultural formations, but rather as composed of culturally constituted processes of communicative interaction" (Mische 2003, 258).

In their incorporation of forms of meaning, these researchers have mainly drawn on three strands in the sociology of culture. As exemplified by Margaret Somers, narrative analysis is the main source for the analysis of stories in social relationships (1994). A second strand is Ann Swidler's theory of culture as tool-kit that is used creatively by actors (1986; DiMaggio 1997). But the most important influence was probably Pierre Bourdieu's theory of social structure as patterns of symbolic practice ([1980] 1990; DiMaggio 1979; Breiger 2000; Mohr 2013).

Interestingly, a number of similar attempts to wed the cultural and the structural seem not to have had an impact on American network analysis. Norbert Elias already conceives of social structures as "configurations" with both structural and cultural aspects ([1970] 2012; Elias and Scotson [1965] 1994). Similarly, symbolic interactionists argue that symbolic meaning evolves and is reproduced in social interaction. Early authors such as Tamotsu Shibutani (1955) and Herbert Blumer (1969) deploy the group concept rather than write about networks (see section 3.1). However, their perspective is easily reconcilable with social network analysis, as Robert Stebbins, Gary Alan Fine, and Sherryl Kleinman have pointed out. They propose that social networks should be seen as "subjective constructs" based on social relationships as dyadic "definitions of the situation" (Stebbins 1969, 11; Fine and Kleinman 1983, 97ff; Salvini 2010). Relational sociologists Nick Crossley, Mustafa Emirbayer, and John Levi Martin follow this lead and conceptualize the subjective and intersubjective construction of networks in more detail (see sections 1.4 and 7.6).

Diverse theoretical strands thus view social networks as interwoven with meaning and call for an incorporation of cultural aspects in network research. But what exactly can we assert about this interweaving of networks and meaning? The remainder of this chapter explores the "meaning structure" of social networks, starting with the symbolic construction of dyadic relationships. The following chapters consider the interplay of networks with wider sociocultural patterns.

2.4. Relationships as expectations

Social networks consist of dyadic ties between individual and/or collective actors, and their coupling in a network structure. But what a tie is, and how it evolves, is rarely discussed in network research. Paul Holland and Samuel Leinhardt explicate the usual approach:

> For our purposes, relations are taken as givens. That is, we are not particularly interested in investigating the substantive meaning of relations. We are satisfied to assume that their meaning and their significance are empirically based. (1977, 387)

In this section, I look behind this analytical approach of taking relations as givens. The obvious starting point is Georg Simmel's formal sociology. Simmel views dyadic relationships as the basic "forms" of society. These forms can be filled with different meaning, according to the thoughts and sentiments of the people involved. But a lot of social dynamics result from the forms themselves (1909, [1908] 1950). According to Simmel, it is primarily the interaction between people, rather than their psychic processes, that determines how individuals are tied to each other. However, his discussion of social forms such as subordination, conflict, or the "tertius gaudens" is not quite clear about whether these forms are ideal types of structure, or whether they follow available cultural models for relating.

A more precise definition of a social relationship can be found in Max Weber's work:

> The term "social relationship" will be used to denote the behavior of a plurality of actors insofar as, in its meaningful content, the action of each takes account of that of the others and is oriented in these terms. The social relationship thus consists entirely and exclusively in the existence of a probability that there will be a meaningful course of social action—irrespective, for the time being, of the basis for this probability. ([1922] 1978, 26f)

Weber's concept of social relationship covers all sorts of social structures, including ethnic groups, formal organization (bureaucracy), political parties, churches, and the state. Here I only apply the concept to dyadic social relationships between two actors. Weber defines social relationships as the *probability* of specific actions between actors. In the first part of the definition, he relates it to the *subjective orientation* of actors toward others. According to Weber and most of the theory of action that follows, social structures exist in the minds of the people involved (see section 2.2). They are structures of subjective meaning, which make certain actions probable and others improbable. Weber's definition thus formulates, in different words, the interplay at the heart of figure 2.1: the systematic probability for social action equals the observable patterns of communication; the relationship itself consists of meaningful orientations making for this systematic probability / the observable patterns. John Levi Martin terms these patterns the "action profiles" characterizing the social relationship (2009, 11). These do not consist of the same kinds of communication over and over again, but of a specific mix of activities occurring and expected in the relationship.

In contrast to Weber, his contemporary Leopold von Wiese views social relationships as *processes*, rather than relatively stable structures (1932, 53ff). Relationships consist in the dynamics of coming closer or of increasing distance between people. Are these two conceptions of social relationships as structure or process antithetical and irreconcilable? Returning to the interplay of regularities of communication and the meaning structure of social networks, we see process and structure as interwoven. The expectations in social relationships make for a certain predictability and regularity of communication. In turn, communication tends to confirm and perpetuate these expectations, but it can also lead to their change.

Importantly, social relationships have to come from somewhere, and that is through the process of *communication*. Following Talcott Parsons's and Niklas Luhmann's concept of "double contingency of action," people build up expectations about each other's actions, and about the others' expectations concerning one's own actions, in the course of communication (Luhmann [1984] 1995, 103ff; Parsons [1968] 1977, 167ff; see section 8.3). These interpersonally negotiated expectations are relatively stable constraints for action in any single situation. Relationships thus reduce the fundamental uncertainty between alter and ego (Berger 1988; Azarian 2010). But the mutual attuning of expectations makes for the incremental change of social relationships over time. I elaborate on this gradual buildup of expectations in chapter 8.

For Weber, Parsons, and Luhmann, all social structures consist of expectations, making for a relative predictability of action/communication, as in the Weber quote. How do social relationships differ from other structures? In social fields like the economy, law, and state bureaucracy, expectations pertain to all actors indiscriminately, at least in principle. All potential buyers have to meet the expectation of paying a particular price for a product; all subjects have to conform to legal norms—and face similar sanctions if not; all citizens have to fill in the same application forms and meet the same criteria when applying for social security benefits or for a passport. In formal organizations, expectations are directed toward positions, not toward individuals. When the general manager or the lowly employee of a company change, the new manager or the new employee still have to perform the same duties as their predecessors. Then there are expectations tied to the individual: I expect the next book by

Hilary Mantel and the next movie by Christopher Nolan to be as brilliant as the last ones.

Social relationships are about two particular individuals, but they concern their behavior toward each other. Neither general brilliance, nor idiocy, nor role-conforming behavior is expected from them. Rather, alter and ego are expected to behave in a distinct way in relation to each other. In a sense, the behavior in a social relationship is "deviant"—it deviates from norms of general interaction and involves particular norms for the communication between the actors involved (Denzin 1970). I expect my wife (cognitively and normatively) to treat me differently than she treats her friends, colleagues, or casual acquaintances. This expectation guides my own behavior toward her, and it makes for a recognizable difference between our communication and that with others.

Social relationships as *dynamically evolving bundles of interpersonal expectations* are obviously laden with meaning. At their core lies a *relational definition of the situation*, a definition of how alter and ego relate to each other (McCall 1970, 11). For example, when a relationship is defined (in the course of its dyadic transactions) as "love," this entails very different expectations than a relationship definition of "friendship," "enmity," or "extramarital affair." As I will argue, communication frequently builds on such "relationship frames" to define the relationship between alter and ego, to establish particular expectations between them, and to reduce social uncertainty (see section 2.7). These form part of the wider culture that leaves its imprint on social networks. However, relationships are free to deviate from these cultural models. They always develop their own idiosyncratic sets of expectations governing the communication between alter and ego.

> Definition 2.4. Social relationships are bundles of relational expectations about the behavior of two actors toward each other. These expectations form part of the "meaning structure" of social networks.

When network researchers mark a relationship as existing (1) or nonexistent (0), they really reduce this complex set of expectations, and the communication taking place, to one bit of information. Whether this is analytically sound has to be questioned with regard to the measurement of relationships and to the specific research question. As Kitts argues, the different measures of ties tackle different aspects of social ties (2014, 271f, 274). Expectations and communication patterns affect each other, but they do not necessarily match, and they are rarely as dichotomous as the prevalent network measures suggest.

Now the notion of expectations seems to imply that the meaning of a relationship lies on the subjective level. In chapter 8, I argue that we should conceptualize expectations as inscribed in communication, rather than view them as rooted in people's minds. Even social psychologists note that the meaning of a relationship evolves in the course of communication within the relationship (Wood 1982; Duck 1990, 25; Wood and Duck 2006). Relationships unfold and develop in communication. To admire or secretly love somebody is not enough to develop a love relationship—not even if the affection is mutual. "Love" has to be communicated and agreed upon between alter and ego. Structures of meaning, including relational definitions of the situation, evolve in communication. In turn, they shape communication by providing expectations about future events. This is the interplay of patterns of communication (level 2) and meaning structure (level 3) as depicted in Figure 2.1.

In a different formulation, Harrison White, Margaret Somers, and Charles Tilly argue that social relationships consist of "stories" relating "identities" (White 1992, 166ff, 196ff; Somers 1994; Tilly 2002, 8ff, 26ff). The term "*story*" emphasizes the historic nature of relationships—they evolve over time in the course of transactions, by building up, modifying, and discarding interpersonal expectations. But it also refers to a narrative approach: personal relationships are shaped by the self-descriptions as narrated within the relationships and to others (Somers 1994; Bochner, Ellis, and Tilmann-Healy 2000; Swidler 2001). I see such stories as affecting social relationships in their buildup of relational expectations. Stories represent expectations, and they are an important vehicle for rendering them explicit and negotiating them. This holds in particular for intimate relationships with frequent exchanges of the form: "But you said that . . ."; "No, I never meant . . ."; "Well, last week you . . ." But relational expectations can exist, and change, without storytelling taking place—simply by the observation of relational behavior between them. If Eric brings Sue roses, this evokes relational expectations without anybody telling the story. However, stories as a form of meaning are again located in communication, rather than in people's minds. Here I agree with White, Somers, and Tilly: communication, rather than subjective dispositions, makes relationships.

2.5. Processes linking multiple ties

Social relationships, then, are inherently laden with meaning as bundles of expectations. As such, they can be seen as relatively autonomous, self-reproducing entities: the communication in a tie gives rise to expectations that mostly conform to these expectations. However, networks are *patterns* of relationships, not their mere aggregation. How are relationships coupled to each other? This section addresses the *processes of communication* linking multiple ties to a network. The following sections then take a closer look at the forms of meaning arising from these processes, organizing them.

First, *multiple relationships* can be *activated simultaneously*. For example, with five colleagues having lunch together in informal conversation, ten distinct relationships are at play (A–B, A–C, A–D, A–E, B–C, B–D, B–E, C–D, C–E, and D–E). Communication then has to take 10 different bundles of relational expectations into account and potentially has repercussions for all of them. This sounds quite complicated, and it is. Often few dyads dominate the conversation, or the discussion turns to safe (and often superficial) topics like sports, movies, or absent colleagues. As Fine and Kleinman note, the simultaneous activation of ties is very important in groups like gangs, friendship cliques, or families (1983, 98f), requiring and making for an attuning of processes and of expectations in the ties at play.

Secondly, a *tie* can become *relevant* and be referred to *in the communication in another tie*. For instance, A turns down an invitation for drinks with his colleague B because he has to pick up his daughter C from kindergarten. The two ties A–B and A–C here remain separate in terms of the communication taking place. But the tie A–C affects the communication in tie A–B through the expectations in it. To take a second example, D blocks her attractive acquaintance E's flirtatious advances because of her romantic involvement with F. This might become explicit in communication, with D saying: "You know that I'm married." Even the casual display of a wedding ring is an explicit reference to relational expectations. But the effect can also remain implicit, in the unreciprocated flirting. In both cases, the second tie influences the patterns of communication in the first tie. Importantly, the second tie does not have an impact on the first without somehow affecting the communication in it.

Both mechanisms start in the process of communication, either in the simultaneity of multiple ties involved, or in the effects of one tie on communication in another. And they lead to patterns of meaning capturing these operational connections between ties. The expectations in a tie then include expectations about other ties around it, most importantly those of the two actors to third actors. In principle, the same holds for other social contexts—such as organizational involvement. The marriage tie "learns" about the spouses' jobs as professor and schoolteacher, and how they affect the communication between them ("You are always working late and not coming home for dinner!"). But such other contexts do not concern us here. The prime form of meaning capturing and coordinating these multiple involvements in ties (and in other contexts) is the construction of identities, as I elaborate in the next section.

Harrison White offers the notion of "switchings" for processes such as the simultaneous activation of ties and the referencing of one tie in another (Godart and White 2010; see section 7.8). Communication here switches between network contexts and between the forms of meaning connected to them (White's "domains"). These switchings lead to "fresh meaning," that is, changes in network domains that otherwise tend to reproduce themselves. However, White does not detail the precise processes at play, or whether communication ever does not switch and make for change in patterns of meaning. The formulations here specify White's idea of switchings in terms of communication processing in multiple ties and leading to expectations fitting and structuring these processes.

2.6. Identities

The construction of *identities* is the most important, and the most basic, form of meaning coupling multiple network ties to each other. The concept is quite complicated, and it differs depending on which theoretical approaches we consult. Harrison White alone discusses five meanings of identities, from the structural position in a network, through a network node's internal reflection of its various entanglements, to the full-blown social construction of a "person" (2008, 17f). I try to keep things simpler. The notion of identity here marks a projection point in the social world to which we attach meaning and expectations. In section 8.5 I replace the concept of identity with that of "actor" in a very similar sense—as meaningful construction, but with a more precise formulation of this construction. For now, "identity" will do.

On the *social* level, identity is the "face" of actors visible to others (Goffman [1955] 1967). Identity here is a social construction that evolves in communication with others (McCall and Simmons [1966] 1978). For example, actors may engage in impression management to convey a certain image to others—but they may also fail in this (Goffman [1955] 1967; [1959] 1990). In the academic field, whether a researcher is regarded as a star professor or an outsider is the result of communication within the academic discourse, and not necessarily based on innate qualities of the researcher (Merton 1968; Collins 1998). For example, German philosopher Arthur Schopenhauer was mostly ignored for more than 30 years after the 1818 publication of his opus magnum *The World as Will and Representation* but rose to fame from the 1850s onward—after shifts in the field of philosophy.

The two notions of "star professor" and "outsider" already suggest network patterns correlating with these identities. In both instances, the communication with the star professor or with the outsider is affected by their ties to others. Knowing about the many admirers and followers of the star professor, I will approach her at the conference very differently than the academic outsider. Similarly, both the star professor and the outsider, aware of their identities, will treat me differently. Whether or not a tie between them and me develops, and what kind of tie, is strongly influenced by their network positions (and by mine). These identities map our network positions, and they coordinate the communication in them. The processes in any one relationship thus have an impact on those in others. My entanglement in one relationship (say, a marriage, or the mentorship of the star professor) bears an imprint on the social construction of my identity, and it can be observed and drawn upon in other relationships.

Over time, dissonant identity constructions tend to wither and result in relatively balanced network constellations. For example, the sole follower of an academic outsider probably changes her allegiance (or drops out of academia) and at least pays respect to the widely accepted star professor. Of course, it might still be possible to separate different social identities if they are left invisible to certain network segments, or if they are seen as irrelevant. For example, a star systems theorist will be ignored by adherents of rational choice, and the academic outsider can still excel at his local soccer scrimmage. Or an unfaithful wife might be able to keep her romantic adventures hidden from her family. These exceptions require specific conditions of secrecy or meaningful separation to not fall under the general law: that the carefully negotiated identity in one social relationship affects the identity construction in others. Identities couple network ties. This is particularly true in network contexts (such as the family, a company, or academia) where the construction of an identity in one relationship can be observed by others and directly affects relationships with them.

> Conjecture 2.1. The construction of an actor's identity couples dyadic ties to a network to the extent that the identity in one relationship is relevant for other relationships.

On the *subjective* level, identity refers to an understanding of what I am in relation to others (Taylor 1989, 27). As symbolic interactionism argues, we develop our subjective meaning and our "self" in our relationship with significant others (Cooley [1902] 1964, 168ff; Mead [1934] 1967, 135ff). This assertion has been confirmed in empirical research (Yeung and Martin 2003). In general, close intimate relationships matter more for our attitudes and for the formation of our identities (Berger and Kellner 1964; Jamieson 1998). The more intimate a relationship, the more elaborate and idiosyncratic its culture, and the stronger its impact on our worldview and our understanding of ourselves. "Secondary" or relatively "superficial" encounters—for example, with the cashier at the local supermarket—are much more routinized and prescribed by "cultural blueprints" for interpersonal behavior than, say, a close friendship that has gone through a considerable amount of negotiation and attuning.

Not all nodes in social networks are individuals. Relational expectations can attach to corporate actors (companies, universities) as well as to collectives (street gangs, social movements; see section 8.9). Just like individuals, they have to establish and negotiate their identities in relation to other actors (other companies, universities, street gangs, political actors). And they similarly struggle to define themselves in the internal processing of meaning, here in the communication between employees, professors and students, gang members, or activists. Since this takes place in communication rather than in thoughts, the internal construction of collective or corporate identities lends itself better to sociological observation than that of individuals.

> Definition 2.5. The identity of an actor consists of the meaningful construction of a coherent, acting entity. This construction takes place within the actor (subjectively for individuals, in internal communication for collective and corporate actors) and in her/his/its social relationships with others.

These different renderings of identity internally and in different relationships (and social contexts) need not match and probably never do. I have argued earlier that the construction of a social identity in relationships will attune and balance, as long as these lie in the same network context. How do we know whether two social relationships involving one particular individual (and her identity) belong to the same network context? Here we can refer to White's notion of "network domain" of a network interwoven with, and integrated by, cultural forms, such as social categories, institutions, specialized language, etc. (Mische and White 1998, 702ff; see sections 2.2 and 2.8). Network ties within the family, in a sports club, in an academic field, or the like, would obviously have to synchronize more than those an individual has across these different domains.

For individual actors, King-To Yeung and John Levi Martin show the subjective and the social level of identity to be intertwined (2003). Identity processes are an important site for the coupling of psychic and social processes. We can hypothesize the same for collective and corporate actors: the more a street gang is held together symbolically by the construction of a collective identity in internal communication, the more it will be able to act in coordination vis-à-vis other street gangs. Also, the interaction with other gangs (and with law enforcement, etc.) will have an impact on the collective identity (see section 8.9). As for individuals, the construction of an identity within a social entity interdepends with that in the relationships with other social entities. The construction of identities not only couples network ties to each other but also makes for an attuning of processes inside and between actors.

Conjecture 2.2. The processing of meaning inside the actor (subjectively or in internal communication) is in interplay with that in the relationships with others in the construction of the actor's identity.

2.7. Relationship frames

The relational expectations between the actors involved and their identities together form the "meaning structure" of a social network (level 3). Both arise from the communication between those involved (level 2), but they also draw on institutionalized "culture" that provides models (level 5). In the case of identities, this includes social categories (like ethnicity and gender; see chapters 4 and 6) and institutionalized roles (like professor and student or husband and wife; see chapter 5). Social relationships are similarly imprinted by cultural models. These "relationship frames" like love, friendship, and patronage are the subject of this section. This includes a brief discussion of the role categories (husband, wife, friend, patron, client) connected to relationship frames (section 2.7.d).

(a) Frames and relational work

As I noted, the relational expectations differ profoundly if the social relationships are "friendships," "love," "patronage," or "competition." What does it mean that two actors have a friendship, or a love relation? I suggest that these are "cultural blueprints" for the organization of a relationship, implying very different interpersonal expectations (Bell and Healey 1992, 308f). While the definition of a relationship as "love" is a feature of the "meaning structure" of the network, the cultural model "love" itself is relatively independent of its deployment in any particular ties. It forms part of the wider culture in the sense of section 2.2, and of the "cultural patterns" in figure 2.1. As I argue in chapter 5, such models or blueprints for relationships are "relational institutions." In modern society, relationship models are portrayed and diffused in the mass media, from Shakespeare's *Romeo and Juliet* to today's blockbusters, novels, and TV series (see section 6.2 for the example of romantic love). The widespread diffusion and availability make for their relative independence from any particular social network.

Following this argument, "friendship," "love," "patronage," "collegiality," "mentorship," "competition," "conflict," "fandom," etc., are "relationship frames"—cultural models or blueprints for bundles of relational expectations (Fuhse 2013, 191ff). The notion of "frames" builds on Erving Goffman's concept of interpretive frames that people, and academic observers, use to make sense of "what is going on" in any particular situation ([1974] 1986, 10f). Cultural frames provide ready-made models for definitions of the situation, such as "play" or "joke" (Bateson 1972, 177ff). As these examples suggest, frames are not only used to decipher communication, but also to structure it. A joke is usually made with the frame "joke" in mind.

Goffman's (and Bateson's) concept of frame has been picked up for the symbolic construction of problems in social movements (Snow et al. 1986), for cultural differences in linguistic expectations (Tannen 1993), and for cognitive models of situations (Lindenberg and Frey 1993, 199ff). Paul McLean examines how Florentine patricians evoke vocabularies of "friendship," "family," "virtue," or "honor" in patronage-seeking letters (1998, 2007). With such keywords, the senders offer to frame the relationships to the letters' receivers in a particular way. Different vocabularies signal varying degrees of loyalty and deference.

In a similar vein, Viviana Zelizer describes this categorizing of relationships into culturally available frames as "relational work" (2005, 33ff; Bandelj 2012). Categories of relationships like friendship, companionship, love, patronage, or mentorship are connected to particular obligations of the actors involved. As Zelizer shows, these obligations extend into the economic and legal realm (2005). But mostly, they pertain to mundane day-to-day behavior. People in a love relationship, perhaps even married, are expected to care for each other, to support each other in times of need, to listen to each other's sorrows, to regularly engage in bodily intimacy, and so on. These kinds of communication form part of the "action profile" (Martin 2009, 11) tied to the respective relationship frames. Adopting such relationship frames reduces the inherent uncertainty between alter and ego. The communication between them can now draw on the cultural model to negotiate and establish relational expectations. Of course, the precise bundle of expectations in relationship frames differs through history, and across geographical and sociocultural contexts (Jamieson 1998; Yeung 2005).

As cultural models, relationship frames need names like "love," "friendship," or "mentorship." Otherwise, it would not be possible to invoke them in communication. However, Zelizer's examples show that much negotiation of the framing in a relationship takes place implicitly. People engage in the practices and use the symbols typically associated with particular frames. In the sense of Goffman, these practices and symbols serve as "clues" or "keys" for how to define the situation ([1974] 1986, 40ff). In the context of relationship frames, they signal commitment to a type of relationship. For example, the "bottle girls" studied by Ashley Mears frame their servicing in the VIP areas of exclusive nightclubs as leisure (2015). They vehemently reject the idea that they are "working" and should or even might get paid for it. Instead, they perform their valuable service as relational work for the "friendships" with the promoters, who in turn invite them for dinner and drinks, give them expensive gifts, or even pay for their apartment as part of the friendship bargain. In spite of all the monetary exchanges around them (and facilitated by their work), getting paid would violate the friendship frame.

Following C. Wright Mills, relationship frames are "vocabularies of motive"—culturally available ways for decoding and interpreting behavior with regard to underlying subjective dispositions (1940). Here, the dispositions pertain to the kind of relationship the actors want to have. Both alter and ego know that their behavior will be interpreted in this regard. This propels them to bring flowers to their lovers, help their friends move, and attend dreaded family gatherings (often at considerable expense). All of these activities are expected in certain kinds of relationships. Through their performance, the relationship frames are enacted and validated for the relationships at play, and it renders the people seen as conforming to the respective roles ("lover," "friend," "family member"; see further discussion). Of course, these practices and symbols can become contentious, as when ego rejects the expensive ring offered to her by alter. Thereby she shows that she does not see their tie in the romantic way proposed.

> Definition 2.6. A relationship frame is a cultural model for (kinds of) social relationships. Relationship frames feature a name ("love," "friendship," and the like) and a bundle of typical relational expectations.

Even when agreed upon, relationship frames do not entirely prescribe what is going on in a relationship. Relationships can switch frames, modify them, or explore the scope of any given frame. In terms of figure 2.1, the culturally available relationship frames (level 5) will be adopted, and adapted, between the actors involved to define their relationships as part of the meaning structure of the network (level 3) to structure the communication between them (level 2). As a consequence, the framing of a particular relationship is continuously in flux. Frames are routinely elaborated and modified to define any particular relationship. In network research, we frequently categorize a number of ties in a network indiscriminately as "friendship." This is starkly simplifying, but probably not too misleading as a rough distinction between different "types of tie"—given that the participants and the communication between them deploy these frames to structure relationships. However, we have to make sure that our classifications correspond to the relationship frames at play, and that these frames are not used very differently by the actors.

(b) Network structure

Relationship frames pertain not only to the behavior between the two actors related through them. They also carry implications for the network of ties around alter and ego. Let me go through a few examples:

- Framing a relationship as *"romantic love"* implies a certain long-term commitment to intimacy. The prevailing ideal of romantic love is heterophilous with regard to gender (Butler [1990] 2007), while being homophilous on many other traits: age, ethnicity, education, lifestyle, etc. (see sections 6.7 and 6.8). Also, it demands exclusivity: Actors are expected to form only one love tie. Earlier ideals of lifelong fidelity (still visible in the marriage frame) have given way to the principle of "serial monogamy," which allows for changing partners, but having only one partner at any time. Structurally, there seems to be a norm against 4-cycles: Even high-school students tend not to hook up with ex-lovers of their former lovers' partners (Bearman, Moody, and Stovel 2004, 73ff). "Spanning tree" network structures ensue, with many branches off a main chain of sexually proactive students. Across types of tie, transitivity kicks in: romantic partners are supposed to become part of their respective families and friendship circles.
- Unlike love, *friendship* ties are still mostly homophilous with regard to gender, and to many other traits and attitudes (Lazarsfeld and Merton 1954; Jamieson 1998, 93ff). It demands a certain degree of transitivity in that friends are supposed to treat each other's friends amicably (Adams and Allan 1998). Network structures with high degrees of local clustering result (Martin 2009, 42ff). However, friendship is a relatively fuzzy concept allowing for all sorts of communication patterns. It seems to be a residual category of important personal relationships not falling under more specialized rubrics (Fischer 1982b).
- *Patronage* links actors in an unequal exchange of services, support, and loyalty (Wolf 1966, 16f; Eisenstadt and Roniger 1980, 49ff). Structurally, multiple clients connect exclusively to one patron, offering their loyalty, political support, information, and sometimes money in exchange for economic aid and protection (McLean 2007). To a certain extent, this reflects a preferential attachment process, since patrons become more powerful and more attractive with the number of clients. However, patrons often depend themselves on patrons on a higher level. This makes

for the institutionalization of hierarchy in linked patronage pyramids (Martin 2009, 189ff).

Conjecture 2.4. Relationship frames structure networks by providing expectations for how to relate to the actors connected to relationship partners.

The network constellations corresponding to the three relationship frames "love," "friendship," and "patronage" are depicted in table 2.1. Of course, these are idealized and not conforming to real-world social structures. But relationship patterns are normatively expected to approximate these typical configurations. Love relationships are still widely supposed to be exclusive and heterosexual; friendships tend to form homophilous gender cliques (with norms against opposite-sex friendships and for the friendly treatment of the members of one's clique); and patronage leads to pyramids with multiple levels, and with clients only ever connected to one patron, but these to multiple clients. Note that the expectations tied to patronage and to heterosexual love, at least in its traditional version, are asymmetric, prescribing different behavior for men and women, and for clients and patrons. Of course, the precise expectations tied to any one relationship frame change over time, and they differ from one network context to the next. That is, the meaning of the frame can change, and with it the network structures corresponding to it (Eisenstadt and Roniger 1980; Silver 1990; Jamieson 1998).

So far, I have portrayed this coordination of network ties by institutionalized relationship frames as if the ties were simply switched on or off by the cultural models. Of course, the coordination has to occur in

Table 2.1. Ideal-typical network constellations by relationship frame

| Love | Friendship | Patronage |

communication. I have already given the example of the flirt between D and E stopped short by mentioning or signaling being married. Of course, E can also learn about D's marriage through other people, especially if they form part of a close-knit community at work or in the neighborhood. Friendship circles frequently meet together, with multiple ties simultaneously activated. This makes it hard *not* to treat my friends' friend amicably. For example, if I go to a bar with a group of colleagues, I cannot well buy drinks only for the three I like best. And I am expected to welcome the friend my friend brought to my party. Of course, I do not always meet my friends' friends, and then I will not befriend them—whether I want to or not (Small 2017, 153ff). Ties have to be coordinated in communication, or they are not coordinated.

(c) Network mechanisms

Following these arguments, it makes no sense to talk about supposed universals in the structuring of social networks. Networks do not per se follow certain mechanisms like reciprocity, homophily, transitivity, or preferential attachment (Powell et al. 2005, 1139f; Rivera, Soderstrom, and Uzzi 2010; Wimmer and Lewis 2010, 139ff; McFarland et al. 2014, 1090f). They only do so *if*, and to the *extent* that, the relationship frames at play suggest these structural tendencies. This runs counter to the early attempts of network physicists to identify universal network patterns across wildly different social, cultural, biological, and technological networks (Watts 1999; Barabási and Bonabeau 2003). Network structures in the social world depend on the meaning of ties—that is, on the expectations in the kinds of relationships involved. When we investigate network structures according to the network mechanisms underlying their formation, for example, through Exponential Random Graph Models, we learn something about the culturally patterned expectations at play (Bearman, Moody, and Stovel 2004; Gondal and McLean 2013).

To take a rather simple example: Mark Granovetter's theory of the strength of weak ties is based on the idea—taken from balance theory—that strong relationships are transitive (1973, 1362ff; Cartwright and Harary 1956). If A and B have a strong relationship, and B and C similarly, chances are high that A and C will develop a strong relationship, too. This assumption of transitivity of strong ties is reasonable in most cases, but not always. For example,

A is B's spouse, and B and C are having an extended affair. Both the marriage and the affair constitute "strong ties" and should make a strong tie between A and C likely, according to balance theory. The two relationships will probably influence each other. But A and C are unlikely to form a strong and positive relationship between them. B will usually try to conceal his affair with C from A. If A and C meet eventually their relationship will likely be one of rivalry, rather than of friendship (let alone "love").

One might argue that triads of married couples and lovers are unstable constellations. But that depends on the possibility of separating the two dyads in communication and on the level of meaning. Similarly, in patronage, if A and C are clients of B in a mafia network, A and C are not supposed to connect to each other and exchange services and goods directly, instead of through the patron. Rather, they find themselves in structurally equivalent positions toward B (as do A and C in the first example of the married couple and the lover, at least in some respects; Lorrain and White 1971). Granovetter sees transitivity as a universal tendency in all constellations of "strong ties." But the examples show: transitivity does not primarily depend on the strength of ties, but on the meanings associated with them. These meanings are found in the cultural models for relationship formation (frames). Accordingly, we should consider these meanings and cultural patterns in network research, rather than proclaim them superfluous.

Neha Gondal and Paul McLean provide a compelling argument that the meaning of ties is linked to different network mechanisms and structures (2013). They find that personal lending in the core of Florentine patrician families is often reciprocated, transitive, and connected to business ties and membership in a political faction. Credit ties are here interwoven with political patronage and with other commercial activity. Outside of this core, personal lending is sparser and occurs frequently between family members. I concur with their interpretation that credit ties acquire a different meaning in these different parts of the network. However, I would argue that personal lending is not itself a "type of relationship" but forms part of the typical activities associated with other relationship frames. Patricians in financial doldrums turn to richer branches of their family everywhere in the Florentine elite. But the core elite is defined by the preponderance of political alliances and business relationships that also involve personal lending. In this interpretation, Gondal and McLean investigate a case of practices (personal lending) connected to relationship frames with different structural tendencies, and this leads to different network patterns within the Florentine elite.

(d) Role categories

In the two instances of love relationships and patronage ties, network structure results from the cultural patterning of relationships between role categories (see chapter 5). The blockmodel technique assumes that network structure is shaped by the types of tie between actors, and by role categories (White, Boorman, and Breiger 1976). Relationship frames and role categories go hand in hand: the frames of friendship, patronage, and marriage include the assignment of alter and ego to the role categories of "friend," "patron" and "client," and "husband" and "wife." These roles are connected to expectations about how these types of actors behave toward each other.

> Conjecture 2.4. The adoption of a relationship frame relates the actors involved in terms of role categories (corresponding to the frame).

Identities, as defined earlier, feature multiple role categories (linked to the various framings of our relationships), and this requires juggling the expectations in different contexts. In a sense, then, we all have "multi-vocal identities" (Padgett and Ansell 1993, 1260; Padgett and Powell 2012b, 24f), with varying expectations tied to us by social context. The expectations in our ties to multiple other actors *within* one network context, in contrast, are captured in the *role* concept. For example, the wife is expected to behave in a particular way not only to her husband, but also to their children, and to her husband's family and friends.

Role categories and relationship frames couple processes in multiple social relationships by providing an overarching symbolic pattering (see chapter 5). This symbolic imprint builds on cultural models rather than developing from scratch (cf. Leifer 1988). We have to distinguish between (a) the cultural material available for framing ties and for assigning actors to role categories and (b) their actual instantiation through deployment (of cultural models), negotiation, and modification in particular relationships and network constellations. While the first belongs to the realm of *culture* (level 5), the latter forms the *meaning structure* (level 3) of social networks.

2.8. Network culture

(a) From dyads to networks

Culture, as understood here, is not some metaphysical level floating above networks (see section 2.2). Cultural forms are produced and reproduced in social structures like educational systems and the mass media (Hannerz 1992; Fuhse 2018b). Social networks also form an important space of cultural proliferation and evolution (Crossley 2015b; Fuhse 2015a; McLean 2017). The basic mechanism lies in the emergence and the negotiation of meanings in the dyad. Howard L. Becker and Ruth Hill Useem note:

> Symphysis of interaction and participation in joint experiences result in concepts, ideas, habits, and shared memories which to the members are symbolic of the pair. (1942, 16f)

In repeated communication between two people, a *relationship culture* evolves, with "concepts, ideas, habits, and shared memories." For George McCall (1970, 15f, 24) and Julia Wood (1982), social relationships *are* the culture that evolves between two people.[4] Empirical research demonstrates that in friendships and in love relationships, a particular universe of symbolic forms emerges, with special idioms, shared memories, and symbols of the relationship (Baxter 1987; Bell and Healey 1992; Hopper, Knapp, and Scott 1981). We can distinguish this "culture" of a relationship from the relational expectations characterizing it. For example, Sue and Betty develop a shared love for 1980s pop music over a splendid night of listening to Madonna, Michael Jackson, and Prince. This now forms part of their relationship culture, and they clearly influenced each other in this regard. The expectations that they can phone each other to talk about their problems and that they will go to clubs playing music from the 1980s are somehow connected to this relationship culture, but they constitute a different part of the meaning of their friendship. The shared love for Prince and Madonna might even survive

---

[4] While the cultural forms play an important role, I find it useful to confine the notion of social relationships to the (relational) expectations between alter and ego (see earlier discussion). The precise nicknames that alter and ego give each other are less important for their relationship than the expectation that they use particular nicknames. Also, culture can be found elsewhere, but relational expectations only in social ties and networks.

the end of their friendship, only it is not really "shared" in communication anymore.

How do we get from this relationship culture to the culture of a network? How do cultural forms diffuse from singular dyads to the wider network? Again, the two mechanisms of communicative coupling from section 2.5 play out. First, adjacent *ties* in a network are frequently *activated simultaneously*, and this allows for the diffusion of knowledge and symbols from one to others. For example, a child eventually hears that her parents call each other "Pat" and "Lucy," rather than "Mama" and "Papa." She will try these names, potentially adopting them. Cultural forms thus diffuse through the network from tie to tie.

Second, the relationship culture developed in one tie has an impact on the inner *cognitive processes, knowledge, and dispositions* of alter and ego, and this affects their communication with other people. Now, if Sue learns from Betty to love the music of Queen, this can also enter the communication between Sue and her boyfriend Carl. He, too, may become a fan. Or he hates the band's music, and repeated quarrels about the domestic music selection ensue. Generally, we can assume that cultural forms diffuse through preexistent ties; that cultural similarity fosters ties; and that ties with dissonance in attitudes, values, and beliefs tend to disappear (or to adapt culturally). This forms the core of Kathleen Carley's "constructural model" of network dynamics (1986, 1991).

In principle, the two mechanisms of cultural forms diffusing via simultaneous activation in multiple ties, and via mental storage and transport across ties, operate separately. Empirically, the two often play out together. Because of these mechanisms, people in tightly knit networks form similar beliefs and attitudes and have the same information at their avail (Erickson 1988, 101ff).

(b) Groups and network domains

As symbolic interactionists have argued, all meaning is specific to channels of interaction and has to be diffused through these (Shibutani 1955). Harrison White calls this the "domain of a network" (1995a, 1038f; Mische and White 1998, 702ff). Domains encompass:

the perceived array of such signals—story sets, symbols, idioms, registers, grammatical patternings, and accompanying corporeal markers—that characterize a particular specialized field of interaction. (Mische and White 1998, 702)

Densely connected networks thus create their own discourse universes, their "idiocultures" (Fine 1979) through which they differ and distinguish themselves from other network contexts. The concept of culture here marks the relative commonality of symbolic forms in the network context and points to their difference between contexts. The pattern of increased internal connectivity and development of a distinct domain of cultural forms is historically termed "*group*" in sociology (see chapter 3). According to Carley, the stability of groups and of the fault lines between them depends on the distribution and the diffusion of information in the networks within and between groups (1991).

White's concept of "netdom" does not merely reformulate the group concept in terms of networks and the cultural forms they host. Groups are usually built from leisurely informal relationships outside of organizations or social fields like the economy, politics, science, or arts. We would not call the diverse members of a company (managers, personal assistants, accountants, researchers, secretaries, administrative personnel, and janitors) or the researchers studying nanotechnology a group—in spite of their mutual observation and dense interaction. However, they form a "specialized field of interaction"—Mische and White's prerequisite for network domains (see earlier discussion). And this interaction leads to a stock of shared information and attitudes, and to a distinct repertoire of symbolic forms observable as the "culture" of the company, or of nanotechnology research.

When exactly do ties form part of a network domain? I suggest two answers for this: First, network domains tend to be delineated, and integrated, by the forms of meaning at play. For example, family ties are built on the institutions of family and kinship, and on the expectations connected to these cultural models (see section 6.5). The distinction between circles of friends, of colleagues, and of family members is a symbolic boundary demarcating spheres of life supposed to remain relatively independent of each other.

Second, the processes of communication in network ties are frequently activated simultaneously (see section 2.5). This holds true for families, as for networks in a company and among nanotechnology researchers. We can assume that these two mechanisms often play out in tandem, with the simultaneous activation of ties coordinated by designated meanings ("family gathering," "business meeting," "nanotechnology conference"). Conversely, these meanings are negotiated and reproduced in the repeated communication within such frames, in particular through the (meaningfully coordinated) simultaneous activation of ties.

I cannot elaborate on the social construction of these diverse social contexts, from informal groups through formal organizations to fields of society, here. They have in common that increased internal interaction breeds distinct repertoires of shared knowledge, attitudes, and symbolic forms. These characterize and delimit the various network contexts—what White calls the "network domain." Investigating urban communes in the United States, network scholars have shown that these can develop radically different ideas, for example, of relationship frames like "love" (Yeung 2005). Also, the degree of consensus depends on the pattern of social relationships (Martin 2002).

> Conjecture 2.5. Network contexts with increased internal interaction breed and reproduce distinct "cultures" of knowledge, attitudes, and symbolic forms.

Communes are a rare case of relatively isolated network contexts. Most network domains are connected to each other through actors taking part in more than one of them (family, colleagues, friends, neighbors, sports clubs, voluntary associations, etc.). Actors, not social relationships, constitute the "weak ties" between these social contexts (Breiger 1974). The existence of such multiple memberships makes for the relative cultural fluidity and permeability of network domains.

> Conjecture 2.6. Cultural forms diffuse through network ties. Multiple memberships of actors in diverse network contexts make for cultural contact and influence between them.

(c) Structural equivalence

According to Bonnie Erickson, not only close-knit groups, but also structurally equivalent positions show distinct sets of attitudes (1988, 109ff). Structurally equivalent actors in a network occupy similar positions in relation to other actors without necessarily being tied to each other (Lorrain and White 1971; White, Boorman, and Breiger 1976; see section 5.1). Examples of structural equivalence are clients in a patronage network (see section 2.7), or mothers in kinship networks (Lorrain 1975). Clients, or mothers, are comparable in their ties to other role categories—patrons, or fathers and children, respectively. Also, gender seems to be a category of structural equivalence (see section 6.6). Close-knit groups are a special case of structural equivalence.

But if structurally equivalent actors sometimes lack ties among them, where should their cultural similarity come from? They cannot spread knowledge, attitudes, and symbols through the social relationships between them. I see three reasons for their similarity in attitudes, knowledge, and symbols, which intertwine in any empirical instance:

(1) Since structurally equivalent actors occupy a *similar position* in social structure (in relation to similar others), they are likely to share a particular *perspective* on the social world. Thus, secretaries of company managers arrive at similar worldviews, even if they cannot share them much.

(2) Structurally equivalent positions correlate with the assignment of particular *role categories* and of expectations connected to them (see sections 2.7 and 5.1). Similar behavior is culturally expected of mothers in relation to their husbands and children. This symbolic positioning (corresponding to their social structural positioning) leads them to see the world in similar terms.

(3) Structural equivalence concerns the connections in different *types of tie*. Only particular attitudes are shaped, and only particular information is shared between actors depending on the kind of relationship. For example, company managers communicate about certain topics with their secretaries, and about others among each other. This *differential communication* makes for common symbolic repertoires and attitudes of company managers (and of their secretaries)—even if they communicate much more often with their secretaries than with each other. In this vein, Peter Bearman and Paolo Parigi show that social relationships between spouses, colleagues, friends, and kin feature different discussion topics (2004). For example, many husbands report talking about relationship issues with their wives, whereas these wives discuss them with their female friends. In such differentiated communication patterns, gendered knowledge and attitudes can diffuse and reproduce—in spite of the strong ties between women and men (see section 6.8).

What do we know empirically about cultural differences by structural equivalence? The newly arrived "outsiders" in the famous study of conflicts in an English working-class suburb by Elias and Scotson are an instance of structural equivalence without a lot of internal connections ([1965] 1994; see section 4.7). They are united by the symbolic exclusion and repulsion of the "established" traditional inhabitants. Their cultural similarity is reactive, in the sense of constructing a counter-identity to their stigmatization. We do not really know whether they share a wider repertoire of knowledge, attitudes, and symbols. Such outsider groups (or, rather, non-groups) lack the interaction enabling a shared worldview, a positive group identity, and collective action. In this vein, Karl Marx argues that the allotment farmers in 19th-century France were not able to organize as a "class" because of the missing interaction between them ([1852] 2005, 83ff).

Brokers between different groups, or leaders with high levels of centrality in their network segments, constitute instances of structural equivalence contrasting with that of a fragmented out-group. Ann Mische shows in her study of social movement leaders in Brazil that their communication styles differ by their position in or between different social movement organizations (Mische 2008, 51f, 240ff). "Bridging leaders" with a high number of affiliations in multiple sectors (e.g., labor unions, student organizations) are experts in "discursive positioning." "Focused activists" with few organizational memberships engage mostly in "reflective problem solving." "Entrenched leaders" in a number of organizations within one sector specialize in "tactical maneuver." And "explorers" straddling numerous sectors show a proficiency in "exploratory dialogue." The four types of leaders occupy structurally equivalent network positions between social movement organizations. In Mische's study, it seems that the different communication styles result as much from the leader's affiliation profiles (their network positions) as being necessary to maintain these. Also, we are not really dealing with a similarity of knowledge and attitudes, but in ways of interacting in the network.

Omar Lizardo's analysis of General Social Survey data looks at brokers with high numbers of weak ties in their personal networks (2006). They resemble each other in their reported interest in multiple musical genres. According to Lizardo, the brokers' "cultural omnivorousness" allows them to maintain diverse personal connections to disparate network populations. Bonnie Erickson makes a similar argument in her study of cultural resources and networks in a company (1996, 236ff). According to Ronald Burt, brokers have a better chance of developing new ideas—but not necessarily similar ones (2004).

This short overview of empirical findings points to a major difference between the cultural repertoires shared in close-knit groups and by structurally equivalent actors. Close-knit groups share knowledge, attitudes, and symbols due to the exchange and mutual influence in them. The cultural similarity of structurally equivalent actors is more abstract. They resemble each other in their creativity, in their access to multiple cultural genres, in their communication styles, or in their reactions to experienced stigmatization. But they do not share ideas, information, or attitudes. In particular, they lack a symbolic boundary around themselves—a collective identity. Their cultural similarity is part of the sociocultural configuration they are embedded in. This leads to a tentative conjecture:

Conjecture 2.7. Structurally equivalent positions show similar cultural profiles in communication styles and access to cultural genres. Their incumbents do not necessarily share knowledge, attitudes, and symbols (unless they also form a close-knit group).

These short few pages do not suffice to deal with the complex interplay of culture and networks adequately. In particular, we still lack a theoretical understanding of how social relationships and cultural relations (between symbols, concepts, etc.) are intertwined (Roth and Cointet 2010; Basov and Brennecke 2017; Lee and Martin 2018; Fuhse et al. 2020). In any case, social networks are not mere structural patterns, but interwoven with forms of meaning. Consequently, we are dealing with sociocultural configurations. However, social networks and culture do not perfectly match, and it is useful to disentangle the mechanisms making for their correspondence, but also for their relative degrees of freedom.

2.9. Conclusion

To conclude this chapter, I briefly summarize the main arguments. Historically, the level of meaning has often been ignored in network research (see sections 2.1 and 2.3). Usually, the causal connection is made between the ordering principles of opportunity structures and of social categories like

gender and ethnic descent on the one hand, and the empirical matrix of social relations on the hand. On the theoretical plane, there are two levels of
network research linking these two sides: the observable patterns of communication and the meaning structure of social networks. These two constitute
the empirical reality of social networks that we are trying to map in network
matrices and graphs.

The meaning structure consists of "relational expectations" about the behavior of particular actors toward particular others. These expectations are
embodied in dyadic relationships (section 2.4) and in the identities of actors (section 2.6). This "meaning of networks" develops over the course of
communication and channels it in turn. But it also builds on relationship
frames and on role categories (section 2.7). These form part of institutionalized cultural patterns that communication draws on to reduce its inherent uncertainty. However, cultural forms do not float "above" networks.
They are negotiated and diffused in them, and they correlate with positions
in networks (section 2.8). Groups feature distinct cultural repertoires, and
structurally equivalent actors draw on more or less cultural genres as well
as displaying similar communication styles. These two instances of group
structures and of structural equivalence are taken up in chapters 3 and 5.

I have argued that expectations and cultural forms are not primarily
located on the level of subjective meaning (section 2.2). Inner processes
of actors, be they individuals, corporates, or collectives, do play a role and
have to correspond to the communication taking place between them. For
example, identities are constructed in tandem in internal subjective processes and in the communication between actors (section 2.6). But what
is shared and communicated is more important, and this is the primary
source of regularity on the network level. Social networks exist on the
level of social meaning, which crystallizes and evolves in the course of
communication.

Much of the discussion remained on the level of dyadic relationships because they form the basic building blocks of networks. But social networks
are not the mere *sum* of relationships—they are *patterns* of interrelating ties.
Numerous measures of network structure can only be applied on this level of
intersecting ties. Several mechanisms couple social ties to each other:

- Ties are frequently activated simultaneously, in particular at foci of
  activity. Communication then has to take the various identities and

relationships between them into account, leading to a situational attunement of relational expectations across ties.

- The communication in one tie can also refer to the expectations in another tie, leading to expectations about how ties affect each other.
- The identities of actors incorporate their involvement in multiple relationships and network contexts.
- Relationship frames, together with role categories, provide cultural models for network structures (in terms of structural equivalence).
- Cultural forms diffuse from tie to tie, leading to the network phenomena like group cultures and communication styles of brokers.

Only the first mechanism, with its focus on communication patterns, does not directly pertain to the level of meaning. While ties themselves are laden with meaning, their coupling is also based on the construction of individual identities, the diffusion of cultural forms, and the adherence to cultural models. In practice, these various mechanisms do not operate separately. They often together make for the cultural patterning of a particular network, for example, for the social control and relative cultural homogeneity in communities (Elias 1974). Therefore, network mechanisms like reciprocity, transitivity, homophily, and preferential attachment are not as universal as sometimes assumed (see section 2.7.c). The tendency to particular network patterns depends on the kinds of relationships ("types of tie") involved, and on the typical expectations attached to them.

Empirical network research has been mainly concerned with the causality between order principles (mainly opportunity structures and social categories) and network structure. Research on the meaning structure of social networks would have to aim at the mechanisms connecting these levels—in order to enhance our knowledge of the interplay of order principles, patterns of communication, and the meaningful construction of networks. Traditionally, the level of meaning is observed in *qualitative* research aimed at an "understanding" of cultural forms (Blumer 1969, 21ff). Yet, recent research shows that symbolic issues such as identity formation or the application of relationship frames can be addressed with *quantitative* data in network research (McLean 1998; Mützel 2002; Yeung 2005). To do so, network data on ties has to be supplemented with data on cultural forms in communication and/or attributes of actors. Methods for such a combination of relational data with attributes include correspondence analysis and Galois lattices (Breiger 2000). However, it is necessary to explore the processes of how meaning crystallizes and evolves and of how it constrains communication in networks. This task of empirical exploration is often better done with qualitative than with quantitative research (Crossley 2010; Fuhse and Mützel 2011; Domínguez and Hollstein 2014; Bellotti 2015).

3

Groups and Social Boundaries

Until the 1950s, "group" was a core concept in the social sciences. Individuals were thought of as parts of groups, and society as composed of groups. Then, structural functionalism on the one side, and ethnomethodology, conversation analysis, rational choice, and the individual-centered methods of quantitative empirical social research on the other side, gradually took over and redirected the focus on the micro- and the macro-levels. The meso-level of the group was by and large shed out of consideration. Nowadays, the group concept is of secondary importance in sociology. Even the level of interpersonal relationships—once the domain of "groups" is now occupied by another concept: "networks." Why this change from groups to networks? And do we have to declare groups as dead or obsolete?

This chapter builds on the framework from chapter 2 to deconstruct the group concept—and to reconstruct it as a special case of network constellation. This constellation is characterized by dense connections within groups and few ties between them. However, the structural pattern only emerges and reproduces in tandem with corresponding patterns of meaning. Groups are constructed in internal communication as relatively homogeneous collectives, with a clear-cut identity and a social boundary dividing "us" from "them." This symbolic division between the group and its social environment—in particular, other groups—effectively structures the communication in the group, but also to the outside world. Groups are thus not only structural, but also cultural. And they embed socially and symbolically in a wider sociocultural constellation. This means that we can meaningfully speak of groups only for very particular cases of symbolically bounded groups. Otherwise, it is better to term social structures "networks."

I start out by sketching the career of the group concept from the end of the 19th century to the 1960s, in German formal sociology and symbolic interactionism, and its subsequent demise (section 3.1). The next section examines the turn toward networks in social anthropology and American structuralism (section 3.2; see also Fuhse 2006). The following sections lay out a relational sociological theory of groups as network phenomena, focusing on the

Social Networks of Meaning and Communication. Jan Fuhse, Oxford University Press. © Oxford University Press 2022.
DOI: 10.1093/oso/9780190275433.003.0003

development of group cultures (section 3.3), the organization around foci of activity (section 3.4), and the distinction and separation of groups from their social environments through styles and symbolic boundaries (section 3.5).

3.1. Early groupism

The group became a central concept in two strands of classical sociology. German formal sociology introduced the group as a basic feature of social structure (a). Symbolic interactionism picked up on the concept and combined it with the focus on meaning from American pragmatism (b). I discuss the basic arguments of these two approaches before giving a short overview of the proliferation and demise of the group concept from 1950 onward (c).

(a) Formal sociology

German formal sociology used the group concept to argue for the relative independence of the social world from individual minds (studied by psychology), and to establish sociology as an academic discipline in its own right. According to Ludwig Gumplowicz, individuals are imprinted by social groups and act accordingly ([1885] 1999, 148ff). Georg Simmel similarly sees human beings as products of social groups ([1908] 1950, 118ff). For him, groups are "social forms" (or constellations) that develop out of the dynamics and exigencies of "mutual effects" between people ("*Wechselwirkung*"; often translated as "interaction"; Wolff 1950, lxiv). Subjective dispositions, such as affects and sentiments, play a secondary role here. They enter social forms as "contents," but only within the confines set by the constellation of actors (Simmel [1908] 1950, 40ff). Simmel views already dyadic relationships as such a social form, though they are characterized by "immediate" interaction and "intimacy" ([1908] 1950, 128ff). The larger the group, the more "impersonal" it becomes. Individuals have to succumb to the "law of big numbers" ([1908] 1950, 110f). Simmel conceives of groups as supra-personal units that follow the logics of mutual influence as they become larger. At the same time, they allow for less and less individual expression of dispositions and sentiments.

Leopold von Wiese, the preeminent German sociologist of the 1920s, does not follow Simmel in viewing groups as an extension of dyadic relationships.

Rather, groups are composed of dyadic ties—they result from the crystal-lization of relationships (von Wiese 1932, 27ff, 415ff). Von Wiese locates social formations—including groups, but also amorphous and short-lived "masses," and "abstract collectives" like the church or the state—on an emer-gent higher level than relationships. According to von Wiese, groups and other social formations achieve a relative independence from individuals as their subjective imagination detaches from the concrete human beings taking part in them (1932, 490f). The members no longer orient toward each other, but to the idea of the social formation. This orientation to an abstract, depersonalized entity makes it relatively autonomous. The ideas about a group (or other social formation) give it a life of its own.

For this qualitative step from individuals to groups, the relations to other groups are of prime importance. Simmel argues that conflict be-tween groups increases their internal cohesion ([1908] 1964, 87ff; Coser 1956). Antagonisms further the development and crystallization of groups. Conflict thus affects the group's behavior independently of the subjective dispositions of their members. Groups behave toward each other like "indi-viduals" (Simmel [1908] 1950, 257).

Simmel adds an important point about the relation between individuals and groups. Particularly in modernity, people are molded by "cross-cutting social circles" ([1908] 1964, 125ff). The individual is influenced by var-ious groups. And her position at the intersection of multiple circles makes her unique. This duality of persons and groups (Breiger 1974) means that individuals do not follow the imprint of any *one* particular group, espe-cially members in groups with divergent norms and cultures. As a conse-quence, groups have to deal with a certain degree of internal variation, and with cultural influences and structural pressures from other groups through co-memberships. Modern individualism necessarily correlates with, and springs from, a changed structure of social relationships. However, the growing complexity and heterogeneity of group affiliations does not make for more autonomy of the individual, but for more cross-pressures and less predictability.

Overall, German formal sociologists offer the following theoretical arguments about social groups:

- Individuals are imprinted by groups (Gumplowicz, Simmel).
- Groups are constellations and crystallizations of social relationships (von Wiese).

- A group possesses a "life of its own" in relation to its members (von Wiese), particularly in situations of group conflict (Simmel).
- Modern social life is characterized by an increasing complexity of social group memberships, and by overlapping social circles (Simmel).

(b) Symbolic interactionism

The Chicago School, and symbolic interactionism as its theoretical core, picked up on the group concept from formal sociology. However, authors like Charles Horton Cooley, William Thomas, George Herbert Mead, Robert Park, Everett Hughes, Herbert Blumer, and Howard Becker were less interested in the structure of groups, and in the role of individuals in society. They combined German sociology with American pragmatism to ask for the link between interaction and symbolic forms. The embedding of the individual in groups, and the relation between groups, thus became symbolic, rather than structural.

The basic idea is that interaction in the group creates a distinctive symbolic universe: the group culture. This entails "sympathy and mutual identification" (Cooley [1909] 1963, 23), a "web of custom and mutual expectations" (Park 1950, 40), or an "ideology," a system of values, norms, and standards (Znaniecki 1954, 132, 136). Tamotso Shibutani formulates: "common perspectives—common cultures—emerge through participation in common communication channels." (1955, 565) Cultural differences build on the separation of communication channels between groups—and reproduce this separation:

> Special meanings and symbols further accentuate difference and increase social distance from outsiders. In each world there are special norms of conduct, a set of values, a special prestige ladder, characteristic career lines, and a common outlook toward life—a Weltanschauung. (Shibutani 1955, 567; see also Fine 1979)

Florian Znaniecki adds an important criterion: the relations in the group have to be "cooperative" (1939, 805, 807f). Only in collaboration, values crystallize as the common ground of communication, and the "we" of a collective actor forms. Conflict and competition, in contrast, make for a divergence of worldviews and group cultures (Becker 1963).

With these diverse contributions, the symbolic interactionist account of social groups advances two key propositions about the interplay of group structures and cultural differences:

- Groups consist of dense internal cooperative relationships as channels of communication, from which a group culture of symbols, norms, and values emerges.
- Cultural differences between groups are rooted in the separation of social relationships, but they also make for the reproduction of this separation.

(c) Further development

After World War II, the group concept diffused widely in sociology. With the contributions of Robert Bales (1950), George Caspar Homans (1950), Robert Merton (1958, 279ff), Muzafer Sherif (1966), and Theodore Mills ([1967] 1984), groups became a sociological core concept across schools and approaches. Edward Shils diagnosed a "convergence of various trends in American social research toward the study of the primary group" (1951, 68). The rise of structural functionalism and the development of statistical methods of social research contributed to the popularity of the group concept at first. But then the same trends—and the reactions to it—led to its demise:

(1) Talcott Parsons first formulated structural functionalism with small groups in mind (based on working with Bales; Parsons [1974] 1977, 43ff). However, his final theoretical system does not incorporate groups systematically. It focuses on the macro-level of societies and their functional subsystems, with systems of interaction as a secondary component (Parsons [1951] 1964; [1968] 1977).

(2) In reaction to Parsons's grand theory, various micro-level approaches surged in the 1960s (Alexander and Giesen 1987, 26ff). Behavioralism, ethnomethodology, Goffman's dramaturgical approach, and conflict theory all stressed acting individuals as opposed to a harmoniously integrated structure of society. Programmatic statements argued against an "oversocialized conception of man in modern sociology" (Wrong 1961) and wanted to "bring men back in" (Homans 1964). Human beings were not perfectly socialized by their sociocultural

environments. Social systems could never control their agentic impulses. These authors focused on the micro-level of individual actions and of ephemeral "interaction orders" (Goffman 1983). The meso-levels of social relationships and groups were by and large ignored.

(3) The early sociometric and network analytical work of Jacob Moreno, Lloyd Warner, and Paul Lazarsfeld formed part of the methods revolution in the social sciences (Scott 2000, 7ff). Subsequently, the standardized questionnaire and the qualitative interview became the dominant ways of acquiring data. The social world now seems to be populated by individuals as bearers of quantifiable traits, and of subjective meaning to interpret.

(4) This trend to the individual in empirical social research coincides with the 1950s diagnosis of a mass society (Nisbet [1953] 1969; Mills 1956; Shils 1962). Individuals were seen as increasingly isolated, discarded from social formations like groups and communities.

Large-scale social structures (Parsons) and the individual (in various interpretive micro-level interpretive approaches, as well as in the theory of mass society) thus replaced the group as cornerstones of sociological theory. The individual is also the focal point of the prevailing methods of social inquiry. Since the 1970s, the group has been relegated to a peripheral existence. It retained a certain prominence mainly in social psychology, particularly in the Bristol School around Henri Tajfel (1981, 1982; Hogg and Abrams 1988). In its experimental research on social identity, the group is a metaphor for any social distinction, created by as little as a coin flip. Social psychology's groups are not preexistent social constellations, but artificially separated. Tajfel and his followers do not discuss the precise implications of the concept, nor whether it adequately maps the social structure of modernity. Within sociology, Gary Alan Fine picks up on symbolic interactionism in 40 years of in-depth research on the construction of meaning in small groups (2012). He considers small groups as the infrastructure of civil society, thus linking to social movement studies (cf. Eliasoph and Lichterman 2003). However, Fine's work remains an exception rather than a prominent research strand.

The woes of the group concept might not only come from the history of the discipline. To talk about a group carries a number of implications: that we can clearly *distinguish between inside and outside*, in particular with regard to who is part of the group, and who is not; that the group is *by and large homogeneous* inside—in the *traits* defining the group and in the *group culture* of symbols, values, and norms; and that *cooperative and intimate interaction* mostly occurs within these clearly demarcated social formations. Of course, nobody quite claims things to be this way. But we cannot well talk and write about groups as meaningful entities without these assumptions. What if we are unable to find such clear demarcations, and homogeneity within them, in empirical social reality?

Already Simmel sees the individual as the member of a number of cross-cutting groups (see earlier discussion). This makes for cultural influences from the outside coming into groups—through the individuals. This relatively simple model building on groups leads to groups as less bounded and more heterogeneous than the term suggests. Empirical reality might complicate matters further. In this vein, Rogers Brubaker forcefully criticizes approaches of "groupism":

> the tendency to take discrete, bounded groups as basic constituents of so-cial life, [. . .] as if they were internally homogeneous, externally bounded groups, even collective actors with common purposes. (2004, 8)

Formal sociology, symbolic interactionism, and much of the work following them, including the Bristol school of social psychology, are guilty as charged.

The decline of the group concept is as much grounded in its inability to capture the complex social structures of modernity, as in the historical trends sketched previously. The group was too small for the grand theory of society, and too small for methodological individualism of action theory and empirical social research. But the concept was also too imprecise, too coarse for rigorous theorizing and empirical observation. These problems led away from the group, and toward the "network" as the core concept for the analysis of social constellations.

3.2. From groups to networks

Two academic strands were instrumental in developing the concept of social networks from the 1950s onward: British social anthropology (a) and American structuralism (b). In this section, I discuss how the two approaches contrast social networks with the established notion of social groups. In brief, groups are seen as particular constellations of network ties.

"Groupiness" thus becomes a variable that can be studied with the formal analytical methods of network research. In (c), I offer a brief overview of the approaches covered in sections 3.1 and 3.2.

(a) Social anthropology

Anthropology never accorded the group as much importance as sociology up to the 1960s. The main research objects were indigenous tribes. These were supposedly isolated units, thought of as culturally homogeneous and densely connected. Cultural anthropology of Margaret Mead, Clifford Geertz, Mary Douglas, and many others studied the cultural patterns. Social anthropology, for example, in the work of Claude Lévi-Strauss, was mainly concerned with kinship structures and role divisions. The network concept only developed when social anthropology turned to more "advanced" social structures. The earliest studies, by J.A. Barnes (1954) and Elizabeth Bott ([1957] 1971), observe the social relationships in a Norwegian village and family ties in London. Modern social structures differ from tribal societies in their cultural heterogeneity and the complexity of social relations. Consequently, social anthropology devised social structure as composed of interrelated ties without clear boundaries: networks.

Many of the early works discussed this opposition between networks and groups. For example, M.N. Srinivas and André Béteille write in their study of Indian caste structures:

> The distinction between groups and networks is primarily one of boundaries. A group is a bounded unit. A network, on the other hand, ramifies in every direction, and, for all practical purposes, stretches out indefinitely. Further, a group such as a lineage or a sub-caste has an "objective" existence: its boundaries are the same for the "insider" as well as the "outsider." The character of a network, on the other hand, varies from one individual to another. (1964, 166)

Srinivas and Béteille do not oppose the group concept per se. Rather, they see groups as a special case of networks: groups are densely connected networks with a clear symbolic demarcation between inside and outside—as in traditional castes and sub-castes. But the authors observe the dissolution of such clear-cut structural entities, in India as elsewhere. The densely meshed, and almost perfectly closed castes slowly make way for looser networks with ties across social divisions. The network concept allows us to observe, and describe, this change from solid, cohesive, and bounded "group" networks to loose and interrelational networks.

The study by Srinivas and Béteille is no exception in social anthropology. It echoes the basic arguments from many other, more central texts (Bott [1957] 1971, 58; Epstein [1961] 1969, 109; Mitchell 1973, 21). All of them conceptualize groups as particular cases of network patterns. In contrast to symbolic interactionism, early social anthropology shows little concern for the patterns of meaning connected to these constellations. An exception is the work by J. Clyde Mitchell in the 1970s. According to Mitchell, group structures are held together by a system of norms and values; they have clearly defined members, with a role structure among them; and they follow common goals and interests. Groups thus consist of "sets of expectations" between its members. These "norms and rules" make for the persistence of social organization in the group, and they allow it to act as a "corporate body" (Mitchell 1973, 31ff). Mitchell also hints that groups are held together by the meaningful construction of a group identity: the rules of conduct build on (constructed) common goals and interests. And "members" are defined by their adherence to the symbols and values of the group.

Mitchell was one of the key figures in social anthropology. But his tentative formulations about cultural patterns and symbolic constructions remain outliers. The focus was firmly on the structural level of social relationships. Groups are thought of as a special case of cohesive network structures, with little attention to meaning.

(b) American structuralism

American network analysis pursues the same basic approach to develop concise graph theoretical models of cliques and groups (Freeman 1992; Wassermann and Faust 1994, 249ff; Moody and White 2003; Newman and Girvan 2004). Groups are conceptualized, and measured, as subpopulations in a network with a higher level of internal connectivity (density of ties). Identifying such subpopulations makes intuitive sense and seems to be a straightforward task. However, most empirical studies only yield multiply overlapping group structures, rather than clear-cut structural regions— hence the bewildering variety of methods to detect such group structures, from fully connected cliques to less restrictive algorithms, I focus on the theoretical assumptions and implications, rather than discuss methodological details here.

Network analysis gets away from the naive image of social structure as an archipelago of closed and homogeneous group units. Barry Wellman writes:

> Network analysts try not to impose prior assumptions about the "groupiness" of the world. They suspect that few social structures are, in fact, sociometrically bounded. Hence they avoid treating discrete groups and categories as the fundamental building blocks of large-scale social systems. Instead they see the social system as a network of networks, overlapping and interacting in various ways. (1983, 168)

Consequently, groups are no longer treated as absolute entities. Already in the 1960s, social anthropology and gang research discuss "quasi-groups" or "near-groups" (Yablonski 1959; Mayer 1966; Boissevain 1968). These were applied to empirical network phenomena with amorphous structures and a diffuse boundary to the outside. Many loosely connected gangs or friendship groups are relatively organized and persistent. But they lack a clear corporate structure and a definition of membership, as in families or in formal organizations like companies or universities. Yablonski lays out a continuum from amorphous phenomena to perfectly organized units (1959, 108f, 116; Boissevain 1968, 544ff). This is in line with the recently advanced methods that emphasize gradual differentiation, overlaps, and nestedness of groups (Moody and White 2003). "Groupiness" (Wellman) becomes variable. Social phenomena are more "groupy," the more they are internally cohesive, the more they possess a clear and exclusive boundary, and the more they are internally organized. The members of such "groupy" units act as "collective."

James Moody and Douglas White note that group solidarity and cohesion have an "ideational component" (meaning) as well as a "relational component" (network structure; 2003, 104). Their development of a network measure of structural cohesion focuses firmly on the relational component. It makes sense to *define* groups in network terms. But we would also want to know why and how such structurally cohesive network patterns come about, to *hypothesize* about the interplay of the ideational and relational. Formal network analysis, of course, can do little but study network patterns.

Nevertheless, the two branches of social anthropology and formal network analysis advance our thinking about informal social structures further.

The naive notion of groups is shed, in favor of an analytical concept of social networks. This allows formulating four central arguments:

- Groups are special instances of networks, with high internal cohesion, few ties to the outside, and a relatively clear-cut boundary.
- Social phenomena display the group characteristics ("groupiness") in varying degrees: internal cohesion, social boundary, organization, and capacity for coordinated action.
- Social groups frequently change in these characteristics over time, for example, through conflict with other groups.
- Social change in modernity leads away from distinct units (clans, villages, tribes) toward the fluid and overlapping relationship patterns of modernity. These are better termed as "networks" than as "an ensemble of groups."

(c) Overview

British social anthropology and structuralist network research avoid the fundamental error of the group concept, as laid out by formal sociology and symbolic interactionism: They do not take groups to be homogeneous entities with perfect internal connectivity and with clear structural separation from other groups. Instead, network research emphasizes the interrelationality of groups, and their internal differentiation. However, the layer of meaning is mostly bracketed. Groups are seen as densely connected segments within larger networks. Structuralism knows little about group culture, and about the symbolic boundary to the outside, both central features of the symbolic interactionist group concept.

What we need, then, is a reformulated group concept that combines these advances. It should detail groups as particular patterns of relationships (network) and elaborate how these form and persist with the construction of a group culture, including a symbolic boundary of the group (on the level of meaning). These advances do not simply aim for a more complicated group concept. Rather, they allow us to identify group structures empirically and to formulate propositions about their emergence, persistence, and disappearance. The remainder of this chapter lays out such a reformulation of the group concept. I build on the ideas from the approaches surveyed in this chapter and on the *relational sociology* around Harrison White. In line with the general perspective in this book, relational sociology conceptualizes, and examines, the interplay of network relations and patterns of meaning (Pachucki and Breiger 2010; Crossley 2011; Mische 2011; Fuhse 2015b; McLean 2017; see chapter 2).

Roughly, we can classify the four distinct approaches in a two-dimensional space, according to the levels of the social world observed, and to the basic concepts of social structure adopted (figure 3.1). Of course, this overview is idealized and drops a lot of work straddling the divides, such as Fine and Kleinman's symbolic interactionist turn to networks (1983). Nevertheless, with regard to levels of observation, German formal sociology, British social anthropology, and social network analysis adopt a mostly structuralist stance, with the pattern of social relations (*social structure*) taking center stage. In contrast, symbolic interactionism focuses on the level of *meaning*, primarily investigating the symbols in a group and its construction of a boundary. Relational sociology combines these interests and studies both the *structure* of relations and the *meaning* attached to it.

Formal sociology and symbolic interactionism base their considerations of the social world on the *group* concept. Formal sociology focuses more on social relationships and on the relation between groups and the individual. Symbolic interactionism is primarily interested in group cultures. Social anthropology, social network analysis, and relational sociology, in contrast, all conceive of the social world as composed of *networks*. Social anthropology and structuralist network research inquire into the patterns of social relationships at play in diverse phenomena. Relational sociology wants to know why certain network constellations emerge and reproduce. It looks

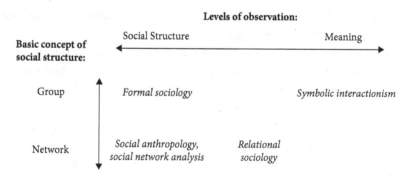

Figure 3.1 Levels of observation and basic concepts of social structure in schools of thought.

for the patterns of meaning developing in network structures and stabilizing them. Following this impetus, my question here is: How, and why do group patterns evolve and persist in the interplay of social networks and meaning? I argue in the following sections that group cultures and symbolic boundaries play an important role.

3.3. Group culture

Following the arguments from structuralist network research, groups can be defined as regions with higher connectivity in a network (see earlier discussion). It makes sense to disentangle the structural and the cultural side of groups when it comes to definitions (Moody and White 2003, 105ff; see earlier discussion). Otherwise, we could not examine how concepts (and the phenomena they denote) relate. Like Moody and White's, my definition focuses on the structural side:

> Definition 3.1. A group is composed of a subpopulation in a network with increased internal connectivity, and with fewer ties to their social environment.

This definition remains on the theoretical plane. It does not specify what exactly "internal connectivity" means, and how it should be measured. But I have to specify that network ties here cover *positive personal relationships* like friendship or love, rather than superficial contact or formal ties, following Znaniecki (see section 3.1). Also, the definition only designates a particular network structure as "group pattern," without detailing where it comes from or how it stabilizes. Its occurrence may be quite coincidental, but with real and systematic consequences.

In line with symbolic interactionism, the repeated interaction in such densely connected groups leads to a particular group culture of values, norms, symbols, and knowledge. I have argued that the following mechanisms make for this emergence of a group culture (section 2.8):

(1) Forms of meaning *diffuse* through neighboring ties. This leads to symbols and knowledge shared in particular network regions, differing from those in distant network regions (Carley 1991; Mark 2003).

(2) Neighboring ties with shared friends or acquaintances are often *activated simultaneously*. The same holds for ties built at foci of activity like sports clubs. In both cases, an attunement in knowledge and values results.

In addition, the density of ties increases cultural similarity:

(3) In densely connected networks, actors face *social pressure* to adopt the values of the group, and to conform to group norms in their behavior.

These mechanisms make for the differential distribution of cultural forms in networks regions. Actors share symbols, knowledge, and worldviews with the people around them—in particular if they are enmeshed in dense networks with high degrees of transitivity. Individual attitudes and values form in social networks and should be seen as their products (Erickson 1988). If, and to the extent that, network regions differ markedly from each other, we speak of each possessing their distinct "culture" in the sense laid out in section 2.2. Harrison White writes of such network cultures as "domains" (1995a, 708ff; Mische and White 1998, 702ff; see section 2.7). These consist of "story sets, symbols, idioms, registers, grammatical patterns, and the accompanying corporeal markers" characteristic of the network they are connected to (Mische and White 1998, 702). Here is conjecture section 2.5 again:

Network contexts with increased internal interaction breed and reproduce distinct "cultures" of knowledge, attitudes, and symbolic forms.

Attitudes, beliefs, and values are *subjectively* held forms of meaning. I would locate the "story sets, symbols, idioms, registers, grammatical patterns, and corporeal markers" stressed by Mische and White on the level of *communication*, rather than in people's minds (see section 2.2). The two levels are closely related, since people mostly say the things they think (though usually not everything), and they think along the lines of the communication they are involved in. However, cultural homogeneity in a network is probably even greater in communication than subjectively. When people hold divergent personal beliefs and values, they keep them to themselves in adversary social contexts, conforming to the prevalent symbols and practices (Cowan and Baldassari 2018). Social pressures and contagion only come about through communication, rather than through the people around me *thinking* in a certain way. Therefore, communicated cultural forms are theoretically primary. But methodologically, we can assess consensus and convergence primarily with regard to subjectively held attitudes, values, and beliefs (e.g., Martin 2002).

Let us look at a few empirical findings. Gary Alan Fine identifies an "idioculture" of nicknames, jargon, and practices among Little League baseball teams (1979; 2012, 34ff). Orrin Klapp examines "collective searches for identity" in the 1960s, with distinct fashions, languages, rituals, and worldviews as emergent features in small groups (1969). The early punk musicians studied by Nick Crossley invent their idiosyncratic style in dense local networks of interaction and cooperation (2015a). According to Nina Eliasoph and Paul Lichterman, the interaction in voluntary associations creates distinct worlds of meaning (2003). These groups pick up on culturally available codes, modifying and incorporating them into their worldviews. "Speech norms" are a particular feature of their group cultures—expectations about the linguistic forms used in the group. John Levi Martin finds that urban communes retain distinct sets of beliefs (2002). These "belief systems" correlate with the patterns of personal relationships in the communes: consensus in the group is biggest when it is hierarchically organized. Working with the same data, King-To Yeung shows that communes differ considerably in their understanding of relationship qualities like "love" (2005).

These studies provide evidence for a tendency toward cultural similarity in densely connected groups. In theory, the dual mechanisms of social selection of friends and partners similar in values and beliefs (homophily), and of their mutual pressure and values toward similar attitudes and behavior (contagion) should lead to culturally perfect homogeneous and densely connected network populations. This tendency should even be amplified by transitivity. However, as White states, "Networks do not have boundaries" (1995a, 1039). Even actors in tight-knit groups have some ties to the outside, making for their relative permeability. Following Simmel, people usually belong to more than one group in modern society (see section 3.1).

Also, tie formation and dissolution have a degree of stochastic randomness to them. Sometimes people get along well with people with regard to some attitudes and activities (e.g., sports, fashion), while differing markedly in others (politics, music). And often enough, people with very similar orientations fail to agree on the behavior expected in a friendship or love relationship. Empirical studies show clearly identifiable bounded group formations to be the exception, rather than a prevailing feature of networks of interpersonal relationships (Martin 2009, 42ff). We have to give up on the idea of a society composed of separate groups. According to White, social networks typically resemble the "messy, inhomogeneous" molecules of "polymer gels," rather than perfectly ordered crystalline structures (1992, 70).

3.4. Foci of activity

Given the stochastics of tie formation and dissolution, and the complex overlapping of groups and cultural environments—how can groups as relatively cohesive and culturally homogeneous formations emerge and persist? First, we note that most of the examples given earlier gather around certain foci of activity where the group members meet regularly (Feld 1981). Fine's Little League teams, Crossley's punk musicians, the voluntary associations of Eliasoph and Lichterman, and the urban communes studied by Martin and Yeung are organized around training grounds, music venues, communal residences, and other meeting points. If, and to the extent that, group members regularly meet the same people at these foci of activity, they will form ties among themselves. And these ties are frequently activated simultaneously, enhancing their dense mesh and the diffusion of cultural forms, as elaborated in chapter 2. As a result, they develop a common repertoire of cultural forms (Morrill, Snow, and White 2005). Activity foci as structures of opportunity for contact lead to repeated communication between the people meeting there, and this makes for the emergence of patterns of meaning connecting them.

However, foci of activity are not simple starting points for the formation of groups. First, many people are drawn to bars, music venues, and training grounds, and into communes and voluntary associations, through the friends, family, and colleagues they already have. Second, most foci of activity have particular meanings attached to them, expectations as to what kinds of people to meet there, and what kinds of interactions take place. Punk music venues will be filled with people who like punk. People join communes because they promise a particular way of living, connected to specialized symbols, knowledge, attitudes, and beliefs. The same holds for voluntary associations and for baseball training grounds. People even move into neighborhoods, and join workplaces, not only for economic reasons.

They might already know people living or working there, or they are part of a particular subculture (professional or in terms of lifestyle). Also, people frequently leave sports teams, communes, voluntary associations, or music worlds if they do not form personal relationships there, or if they do not share the prevalent worldviews.

Foci of activity are not independent givens in relation to social networks and group cultures. Rather, social networks and group cultures crystallize around foci of activity, imbuing them with meaning and with personal connections. A complex interplay of structures of opportunity, communication patterns, and meaning structures ensues, rather than one-sided causation. We can formulate in a less determinate fashion:

> Conjecture 3.1. Many cohesive groups, and their cultural repertoires, evolve around foci of activity.

An isolated village would be the prototypical case of a "total" focus of activity, with most personal relationships within, and with few cultural influences from the outside (in the absence of schools and of electronic media of communication). However, we would not call the people living in such a village a "group." The term denotes cohesive sociocultural formations that *could be otherwise*, that have the chance to organize differently. Thus, cities are habitats of a *multitude* of groups (Park and Burgess [1925] 1967, 38, 40ff). But the village does not house *one* group.

The two groups studied by Norbert Elias and John Scotson live together in an English working-class suburb (called "Winston Parva"; [1965] 1994; see section 4.7). They are based in different neighborhoods, but they regularly come in contact with each other. Their structural and cultural separation is organized partly around their different places of residence. In addition, the two groups, through processes of interaction and symbolic construction, come to see each other as different kinds of people. Their stigmatization (by the established residents) and counter-stigmatization (by the newly arrived) erects a boundary of meaning between the two groups. This symbolic distinction governs the interaction between the two groups, rendering cross-group personal ties improbable. And it can even stabilize the residential segregation, when both groups do not aspire to move into the other one's neighborhood, or when housing applications from the wrong side of town get rejected.

3.5. Styles and social boundaries

(a) Social boundaries

Many empirical examples of group formations feature such a symbolic boundary. In the founding statement of the labeling approach, Howard Becker argues that groups of jazz musicians and marijuana smokers develop their very own universes of notions and attitudes *in opposition to the people around them* (jazz audiences, non–drug users; 1963). As in the study of Elias and Scotson, they draw a symbolic boundary that sharply distinguishes between "us" and "them." Crossley's punk musicians in the 1970s created an oppositional culture, against dominant mainstream pop and symphonic rock (2015a). Urban communes explore "alternative" ways of living together (Zablocki 1980). Similar oppositional stances can be found in street gangs (through deviance and violent conflict) and social movements (through counterculture and protest).

In a way, all symbolic forms establish a boundary between something inside, which is designated by the form, and some outside, against which it contrasts (Spencer Brown [1969] 1972). We identify chairs, nations, sociology, and the present in our world, as opposed to tables, classes, economics, and the past. As structural linguists argue, we thus create particular kinds of objects in relation to other kinds of objects. These symbolic relations need not be tied to social divisions, unless they classify kinds of people. Such a *social boundary* isolates a particular group of actors and pits them against others (Becker 1963; Zerubavel 1991, 41ff; Lamont 1992, 9ff; Wimmer 2013, 101ff).

Note that I use the distinction between social and symbolic boundaries differently than Michèle Lamont and Virág Molnár (2002). Their "symbolic boundaries" are all *social* in the sense of marking kinds of people as different. In contrast, I insist that other kinds of boundaries—around social entities such as a university, or between objects such as chairs and tables—are also symbolic; they are meaningful constructions rather than objectively given. In contrast, Lamont and Molnár reserve the term "social boundaries" for categorizations that make for different distributions of resources, such as differences between classes in socioeconomic resources. This constitutes a subset of social categories, since not all of them are necessarily tied to socioeconomic inequality. With my wider usage of the two terms, I follow Andrew

Abbott's use of the term "boundary" (1995) as a symbolic distinction. All social boundaries are symbolic in nature, but not all symbolic boundaries are social in the sense of distinguishing between kinds of people.

According to Abbott, such symbolic boundaries play an important role in organizing all kinds of social structures by constituting "things" in the social world, including professions, groups, and classes, but also universities, companies, and states:

> it is wrong to look for boundaries between preexisting social entities. Rather we should start with boundaries and investigate how people create entities by linking those boundaries into units. We should not look for boundaries of things but for things of boundaries. (1995, 857)

Abbott sees such boundaries as building on "proto-boundaries"—distinctions that are at first only semantic (1995, 867). For example, different kinds of tasks in a company have different terms associated with them—accounting, research and development, etc. In a second step, such semantic proto-boundaries come to distinguish kinds of people (1995, 871f)—they become "social boundaries" in my terminology. By using the terms associated with tasks, we can tell employees charged with performing them from those with other tasks. The people at the company are now divided into researchers, accountants, marketers, cleaning personnel, and so on. The semantic distinction between tasks becomes a social division, organizing the separation of kinds of people and the relationships between them.

> Definition 3.2. A social boundary symbolically separates a group of actors from others around them.

Caste groups are traditionally tied to particular tasks/professions (Sharma 1999). For example, untouchable castes are responsible for dealing with excrement and for shoemaking. Brahmins, in contrast, perform rituals and act as intermediaries to deities. They become teachers and priests. Otherwise, few groups build on particular professions or tasks. Many groups center on activities, values, ideologies, or places. We can hypothesize that all sorts of social practices and symbols can become typified (Abbott's *proto-boundaries*) and come to serve as signs of group membership (as *social boundaries*). Building on the discussion in section 3.3, these are aspects of group culture. In contrast, mental dispositions like attitudes, values, or beliefs cannot, themselves, be group markers. Subjective states and processes remain obscure. Effective signs have to be visible. Therefore, proto-boundaries can refer to symbols and practices, including linguistic features (ways of speaking), rituals, and clothing. Such observable practices can be interpreted as signs of mental dispositions, and they often are. Wearing a kippah or a headscarf visibly marks religious beliefs to others, though the subjective beliefs themselves (or lack thereof) remain hidden from sight.

(b) Group styles

Recognizable practices and symbols constitute the *style* of the group. White defines styles as "packages which combine signals with social pattern" (1992, 166; 1993, 63ff). Behaviors follow a style if falling into distinct "packages," and, if they are recognized as such, effectively signaling social relations of similarity and difference. Here we are only concerned with group styles, that is, patterns of behavior shared in close-knit networks and marking the group membership. Similarly, Nina Eliasoph and Paul Lichterman conceptualize "group styles" as "recurrent patterns of interaction that arise from a group's shared assumptions about what constitutes good or adequate participation in the group setting" (2003, 737). However, their empirical examples concern more the shared cultural repertoire and practices in the group—what I call "group culture" (see section 3.3)—rather than symbols and practices that *signal the difference between the group and its social environment*. In line with the prevalent use of the "style" concept in sociology (Hebdige 1979), group styles should be recognizable not only within the group or by researchers; they also mark the group symbolically in the social world. Individuals exhibiting the group style will be seen as belonging to the group by outsiders.

> Definition 3.3. Group styles consist of symbols and practices marking the membership in a group.

Adolescent inner-city Black males in the United States frequently mark their allegiance to street gangs by "showing colors." For example, the rival Los Angeles–based gangs "The Bloods" and "The Crips" traditionally signal their allegiance by wearing red or blue pieces of clothing (caps, scarves, ribbons), as well as by certain hand signs and graffiti (Monti 1994; Decker and Curry 2002). This combines with other indications of the gang membership ("gangsta" style clothing, particular greetings and other forms of interaction, hip-hop, skin color, gender, age). In combination, this pattern of practices and symbols makes for recognizable group styles, often enough to trigger violent conflict. With these colors (and other appropriate practices), gang members self-proclaim their group membership and make the group a recognizable entity—a "thing" in the sense of Abbott.

To take another example, the punk musicians studied by Crossley developed a repertoire of countercultural forms (fixing pins and chains, spiky hair, ripped jeans, swastikas) and typical musical motifs (instrumentation, chords, rhythm) as elements of their style (Hebdige 1979; Crossley 2015a). This is what made them "punks," and recognizable as such. Also, baseball teams sport distinctive team jerseys and cheers (Fine 1979). All of these stylistic features mark both similarity within the group, and difference to the outside, often to other groups. They are part of the symbolic boundary separating the group from its social environment.

Following conjecture 2.6, close-knit group structures foster particular symbols and practices. Cultural differences, for example, in linguistic forms and interaction styles or in musical preferences, develop in repeated and dense interaction. Such cultural differences can become proto-boundaries through semantic distinction. Then they serve as group styles, marking the inside from the outside.

> Conjecture 3.2. The development of particular cultural forms in close-knit networks makes it possible to mark these groups symbolically with group styles.

The distinction is gradual, but important: A "group culture" of shared symbolic forms develops in the close-knit interaction in the group. A group of friends might influence each other to adopt a liking for wearing leather jackets. However, if leather jackets become a marker of group identity, the meanings attached to these symbols are no longer only negotiated among friends. Rather, they are subject to the interactive construction between the group and the social environment against which it defines itself. The "group boundary" is not a property of the group by itself, but of its interaction with the wider social context. If everybody else also wears leather jackets, they no longer mark group membership.

> Conjecture 3.3. The group boundary is negotiated in the interaction between the group and its environment.

Generally, the wider public rarely reacts to countercultural symbols by adopting them. Sure, the "Gangsta" rap and clothing of American street gangs have diffused into mainstream culture. But the prevalent reaction to songs praising gang violence and sexist gender identities, and to young Black males wearing heavy gold chains and jeans far below the waistline, is revulsion. Elias and Scotson diagnose a cycle of stigmatization and counter-stigmatization between the established residents and the newly arrived in "Winston Parva" ([1965] 1994). According to Fredrik Barth, the identity of a group and its symbolic boundary separating it from the outside are subject to processes of "self-ascription" and "other-ascription" (1969). The two often intertwine in conflict: An esoteric commune may view itself as chosen and enlightened, while regarded as lunatics from the outside. In turn, the commune sees the wider public as blind and doomed. The social boundary is an object of symbolic contestation, with at least two sides influencing and reacting to each other. William Graham Sumner formulates that an "ingroup" always defines itself against an "outgroup" ([1906] 1959, 12f). This out-group could consist of the wider public, and/or of one or more rival groups. In his study of marijuana smokers and of jazz musicians, Becker examines the interplay of a group and the wider public in detail (1963). Street gangs, in contrast, are pitted symbolically both against other gangs, as well as against mainstream institutions like school, the police, and the family.

(c) Boundaries in networks

If, and to the extent that group cultures become recognizable group styles, the social boundary will affect the structural layer of *social networks*. Crips and Bloods are unlikely to befriend each other, in spite of their similarity in linguistic codes, music preference, sociodemographics, socioeconomic situation, and even fashion (apart from their "colors"). Instead, they tend to form relatively cohesive networks among themselves, organized around a core of key members in dense interaction (Klein 1995, 60ff; Papachristos 2013). White introduces, and Stephan Fuchs picks up on, the notion of "involution" for this process (White 1992, 35, 75; Fuchs 2001a, 51, 191f). Involutions crystallize in a network when communication starts to organize around symbolically constructed unity ("collective identity") on the inside, marking the difference to the outside. This makes for an increase in internal cohesion, and a thinning out of positive personal relationships to the environment. Fuchs coins group phenomena as "involuted networks" (2001a, 211ff). The sharp symbolic boundary organizes the interaction within the group and evokes a "sense of belonging." Fuchs's examples for groups are families, cults, and social movements.

> Conjecture 3.4. Social boundaries foster positive ties within the group.
>
> Conjecture 3.5. Social boundaries impede positive personal relationships between a group and its environment, in particular between rival groups.

Conversely, we can hypothesize social boundaries to emerge and persist if they roughly map preexistent network patterns. According to Roger Gould, movement identities have to conform to interpersonal networks in their construction of "us vs. them" (1995, 15ff). Abundant ties across the category invoked make it less convincing, and less able to mobilize. However, the impact of networks on group boundaries might work more indirectly, through the emergence of group culture and its use as a style demarcating the group (conjecture 3.2). With this caveat in mind, we can formulate a tentative conjecture of a direct impact of networks on boundaries, to test in empirical studies disentangling these various factors:

> Conjecture 3.6. Social boundaries are more convincing and salient if they map preexistent social networks.

Figure 3.2 presents a highly stylized hypothetical network graph resulting from these processes. A population of 18 actors is connected by a complex mesh of ties with three relatively cohesive group clusters. These feature a particular group culture of norms, attitudes, worldviews (marked by shaded areas), and social boundaries drawn around them. However, some actors do not fall into one of the groups, and two are located in more than one group (connecting them to each other). Empirically, we would expect everybody to be a member of three or more overlapping groups (friends, family, colleagues, neighbors, etc.). Actual patterns of personal relationships look even more confusing, and less clear-cut than here.

What do social boundaries consist of? Obviously, they feature symbols to mark the difference between in-group and out-group(s). Social boundaries therefore constitute a subclass of symbolic boundaries (see earlier discussion). But symbols alone could never make or break social ties. Charles Tilly argues that social categories are rendered plausible through "stories"

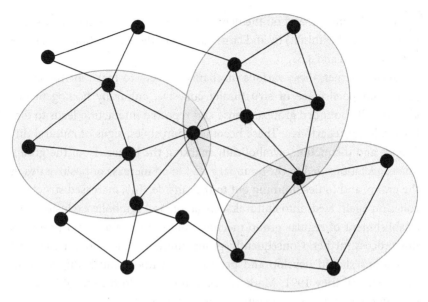

Figure 3.2 Group patterns in networks (stylized).

(1998, 63f). These give accounts of how types of people have connected in the past. They construct the categories as identities in relation to each other. This follows White's assertion that identities are constructed and related in stories (1992, 166ff, 196ff; Somers 1994). I build on this idea but criticize it as too vague (see sections 2.4 and 8.5). Stories can fill categorical identities with life, making them plausible. But like relationships between actors, the relations between groups or categories consist of *expectations*.

Here, the expectations pertain to how members of one group behave toward one another, and how they behave toward members of other groups. In this sense, they are "relational expectations" as opposed to expectations concerning the behavior of isolated individuals ("identities") or about behavior in general ("norms"). As such, they are inscribed in the "meaning structure" of the social network in and around the group (see section 2.1). Contacts within the group are framed as solidary because, as members of one group, they share a common destiny. And contacts to the out-group are supposed to be competitive and conflictual. The communication between actors does not have to follow these expectations. But then they have to face sanctions in their groups. For example, a gang member hanging out with members of the rival gang will have his loyalty questioned. And mistreatment or conflicts within the group could lead to exclusion. Depending on the institutionalization of the social boundary, and on the degree of adversariality, contact across the boundary (and within it) by and large adheres to these expectations.[1] Hence conjectures 3.4 and 3.5.

Taken together, these various mechanisms seem to make for the emergence and persistence of structurally cohesive, culturally homogeneous, symbolically bounded groups. Dense and repeated interaction leads to distinct cultural repertoires. These become group styles, signs of cultural difference and distinction, symbolically marking the boundary of the group. The expectations tied to the boundary lead to an increase of positive ties in the group, and to ties thinning out to the outside. This increased structural cohesion again feeds into cultural segregation and symbolic exclusion. The establishment of regular group meeting points (foci of activity) promotes this process further. Consequently, group cultures, styles, boundaries, and formations should develop and persist even without much difference to begin with (Carley 1991; Mark 1998). Youth subcultures probably spring from this tendency for sociocultural differentiation.

But overall, we do not observe ever-increasing group isolation and conflict. There have to be counter-tendencies. Otherwise, we would probably still speak of "groups" as a central feature of modern society. I have argued that most people are nowadays members of more than one group. And ties form and dissolve stochastically. We can imagine people leaving a gang because of a lost quarrel for status or after forming a romantic tie to a nonmember. They then bring their cultural repertoires into quite different social circles. Also, systems of education and mass media make for an influx of cultural material, challenging the sociocultural universe of the group. Crosscutting social circles, the unpredictability of human attraction, and cultural influences from the outside cause cultural and structural permeability.

(d) Group structure

As a result, the construction of a social boundary can never ensure perfect structural continence and cohesion in a group. The boundary draws a sharp symbolic distinction between in-group and out-group. But the structural level remains interrelational. Most groups differentiate into a densely

---

[1] Like interpersonal relationships, social categories combine normative and cognitive expectations, in the terms of Johan Galtung (1959; see section 7.3).

connected core of leaders, and a periphery of fringe members. These take part in the group activities, frequently coordinated and led by the core. But they retain ties to the outside and identify less with the group. The core defines the group identity and lives it. And the periphery brings irritations from the outside (Fuchs 2001a, 273f, 281ff). Groups conceive of themselves as homogeneous and bounded. The network of social relationships, however, shows only gradual differentiation. Typically, even the distinction between fringe members and sympathetic bystanders remains ambiguous.

> Conjecture 3.7. Social groups are differentiated into a core of leaders and devoted followers, and a periphery of fringe members, based on their respective positions in the network. The core defines the identity of the group (and its boundary). The periphery brings in irritations from the outside.

Through the mechanisms sketched in this chapter, the structural position in the group (core/periphery), participation in group activities, and adherence to the group norms and group style will correlate. A group leader is likely to participate more in group activities than a peripheral member. She will also have more ties in the group, in particular to others in the core. She will identify more with it, signaling this through group insignia. And she is likely to hold the values and worldviews of the group in relatively pure form. The peripheral member, in contrast, might doubt her group allegiance, mix the group style with other cultural forms, and maintain ties to the outside. However, there are exceptions, for example, when a new group member, still located in the periphery, enthusiastically takes part in group activities, forms new ties, and aspires to a more central position. A former leader, in contrast, slowly drifts out of the group after ceasing her group activities.

> Conjecture 3.8. Structural position in the core or periphery of a group, engagement in group activities, identification with the group, and adherence to the group style correlate.

All of these vary by degree, not either-or. Group members are positioned more or less in the core or the periphery; they take part in few, some, most, or all activities; they identify more or less; and they show varying degrees (and variations) of the group style.

The correlation between these variables holds across individuals, but also across ties. Communication in a core tie is more likely to display the group style in pure form, while a peripheral tie can feature doubts and outside influences. What about ties between the core and the periphery, between a leader and her followers? As I argued, all social relationships consist of expectations about how the actors should, and will, behave toward each other. Based on their network positions in the group, the tie between a leader and a follower is likely to be asymmetric. The core member is likely to have higher status and to expect deference. The peripheral member, being a part of the group, will orient her behavior to the expectations signaled by the core. With regard to worldviews, the core member displays firm beliefs, whereas peripheral followers face more doubts and uncertainties, since they have more contact with other social contexts. Some groups also have institutionalized expectations concerning the behavior between members with different roles, e.g., leader and follower (see chapter 5). Overall, core–periphery ties feature a certain imbalance in terms of status, commitment, and cultural imprints.

> Conjecture 3.9. Core members have higher status in the group, making for an asymmetry of interaction and relations with more peripheral members.

A group might try to determine who exactly belongs to it by calling "us or them!" in an effort to close ranks and to seal itself from the outside. But it will usually not be able to enforce a rigid regime of exclusion. Groups need a permeable personal boundary, if only to recruit new members. Old members inevitably drop out as a result of stochastic events and change in personal relationships. Leaders of gangs and terrorist groups are more likely to get killed or to be incarcerated than fringe members, due to their intense participation in illicit group activities. These members and leaders have to be replaced, or the group withers. Alternatively, the group could try to formalize membership, as many voluntary associations do. However, then there are members who pay their fees but do not show up, and others who come to meetings without paying. The same holds for gangs who introduce initiation rites. Informal groupings cannot install a clear distinction between members and nonmembers, since they are not able to provide material incentives, as companies, schools, or public administrations do. Also, groups cannot become too formalized in their structure without losing some of its appeal of authenticity, spontaneity, and creative opportunities.

These difficulties point to the group boundary as *symbolically generalized*. The boundary is not drawn around specific people, but around kinds of people. Through the symbols associated with it (name, insignia, group style), it achieves a relative independence of particular actors. The category "punk" does not signify: Sid, Johnny, Shane, and Souxsie are punks. Instead, it marks people with ripped jeans, needle pins, spiky hair, and a "fuck you" attitude as punks. Similarly, with gangs, it is: everybody who wants to join us wears the right colors and clothing, obeys orders, and engages in gang activities is considered a gang member. The group boundary is relatively general in the expectations tied to it.

> Conjecture 3.10. Groups with symbolically generalized boundaries are better able to cope with change in personnel. Other things being equal, they are more likely to persist.

We do not observe this kind of generalization in friendship groups. Consequently, they are more "network" than "group." They usually do not persist long, and they are less cohesive and structurally separated than group phenomena. Families are also not generalized in their membership criteria. Most people relatively clearly either belong to the family or not. But should the cousin's new boyfriend be treated as family? In any case, people are rarely kicked out of the family if they do not adhere to a common style. The family is a very particular case and cultural model (see section 6.5), which does not really conform to the theory of group relations offered here.

3.6. Alternative accounts

How does the approach presented differ from other contemporary accounts of group phenomena? I focus on a few prominent approaches in and around relational sociology.

First, Pierpaolo Donati and Margaret Archer reserve a prominent place for group phenomena in "*relational realism*" (Donati 2011, 90ff; 2013, 113ff; Donati and Archer 2015, 53ff). Along with dyadic ties and associations, group phenomena are christened as "collective relational subjects" and as a core feature of "relational society." This implies:

- Relationships and groups (and corporate actors) conform to the same theoretical models, partly in accordance with Simmel (see section 3.1). I emphasize the distinctly relational nature of ties: they concern expectations about how two actors will behave toward each other. Groups, in contrast, display a certain abstraction from individuals by virtue of the construction of a social boundary and group identity, notwithstanding the participation of particular individuals. With this "symbolic generalization," groups become gradually more and more independent of individuals (Simmel again).

- By proclaiming groups as social relations, Donati and Archer do not consider the importance of interpersonal relationships in group phenomena (McAdam 1986; Opp and Gern 1993; Gould 1995; Diani and McAdam 2003; Pedahzur and Perliger 2006; Papachristos 2013; Diani 2015).

- True to Donati and Archer's credo of critical realism, groups either exist or not for them. They are "real entities" with causal powers, or mere constructions. My approach examines the conditions for, and the processes of, group construction—if and to the extent that it takes place. Also, groups can display varying degrees of group closure, structural cohesion, and cultural homogeneity. They are not objectively "real" entities, but variable constructions, differing in their "groupness" on three dimensions. Following the critical realist framework, Donati and Archer adhere to "groupism" as derided by Brubaker (2004).

Gary Alan Fine's *interactionist research on groups* is much more empirically grounded than Donati and Archer's theoretical considerations (2012). Therefore, he does not declare them ontic entities but shows that similar mechanisms of the development of idioculture and norms, of status hierarchies, and of role constellations play out in a variety of group phenomena— from youth sports teams, mushroom collectors, the fantasy role-playing scene, and high-school debate clubs to restaurant kitchen staff, self-taught art markets, weather service offices, and political activists. Some of these cases are closer to formal organizations or subcultures, mixing different ideal types of social structure.

However, I see Fine's work as overestimating the role of groups in society, and as treating them as more or less unproblematic units. He writes of society as "an ecology of groups" or of consisting of "a network of group cultures" (2012, 20f), and that groups "constitute and organize political life" (2012, 124). Apart from the overemphasis on groups, these are not considered as existing to varying degrees. Wherever people engage in common activities (whether coordinated or not), they are supposed to form a group. In his most recent book, Fine elaborates on social relationships and networks as the basis of group structures and collective actions (2021). This runs in tandem to my arguments, but he still treats groups as clearly delineated phenomena. While more elaborate and refined, he adheres to a groupist perspective as criticized by Brubaker. Groups are taken to be unproblematic and clearly demarcated units, and they are not broken down into dimensions like cohesion, cultural homogeneity, and group identity that actual phenomena display to varying degrees.

A third prominent approach, by Randall Collins, examines the dynamics and dependencies of group constellations in *interaction ritual chains* (2004). Though he does not use the notion of groups, Collins examines conceptually how constellations of actors stabilize around common practices (like smoking) and mutual orientation. Following Durkheim, he envisages multiple actors as oriented toward a common focus of attention (be it cigarette smoking, or a person), which gets invested with "emotional energy" and becomes a "sacred object" of the group. This concentration of emotional energy makes it rewarding to take part in common activities, and to immerse in the group. The group constellation persists if, and to the extent that, this cycle of investing emotional energy in sacred objects, and drawing from them, remains intact.

The notion of "emotional energy" is obviously an "as if" construction—subjective states and processes, such as emotions and dispositions, can never leave individual minds. A hardcore subjectivist would reject locating emotions in cigarettes. Nevertheless, it can make sense to model things that way ("as if"). Collins's model allows us to see the effective organization of emotions in group phenomena, with an investment in both the group identity and the social boundary to the outside. However, Collins's intuitively plausible description of group phenomena only partly lends itself to empirical observation. Given that emotional states here act as an intermediating factor, leaving them out of the model makes more sense to me—instead modeling group phenomena based on network structures, processes of communication, and the patterns of meaning attached to relations and ties, and inscribed in communicative events taking place.

Last, I briefly address the philosophical debate about "*collective intentionality*" (e.g., Tuomela and Miller 1988; Gilbert 1990; Bratman 1992). Various philosophers here take as their starting point autonomously acting individuals and ask how it is possible for them to "do something together" and to become "collective agents." Again, the crucial steps are located on the subjective level. Most authors take shared intentions or, rather, the shared intention of doing something together, as the decisive step toward collective action. This line of thinking brackets the most important sociological questions and, indeed, their answers. Two or more people wanting to go for a walk together is neither necessary nor sufficient for them to actually walk together. Shared activity requires coordination and communication. Often enough, after communication has taken place, people end up going for a walk together without initially intending to. Subjective dispositions are not the best predictor for coordinated action. We had better look for sociocultural constellations that regularly breed coordinated action. These are social relationships, or groups characterized by social cohesion (necessary to coordinate), group culture (with similar values and attitudes, but also practices and symbols), and a social boundary uniting the group symbolically against other groups, or against the outside world.

3.7. Conclusion

(a) Overview

The historical discussion in sections 3.1 and 3.2 shows that groups are *not* a ubiquitous feature of the social world. Social structure can usefully be detected in patterns of social relationships—networks. Groups are a special instance of network patterns, the exception rather than the rule. One of their core features, and the most easily observable, is a certain degree of *network cohesion*. However, densely integrated networks do not come and solidify out of the blue. I have argued that they gather around foci of activity (section 3.4). In the context of this book, the connection to forms of meaning is more important. Groups as cohesive network patterns persist if, and to the extent that, they are stabilized by the development of a repertoire of shared forms of meaning (group culture; section 3.3), and by the construction of a social boundary separating them from their social environment (section 3.5).

Groups do not become homogeneous and coherent entities through these processes. In modernity, group members are always involved in multiple social contexts, bringing cultural irritations and heterogeneity into any one group (section 3.1). A differentiation between core and periphery in the group ensues (section 3.5). Core members tend to be more connected in the group. They identify, and are identified, more with it, have higher status, and adhere to the group culture in almost pure form. Fringe members, in contrast, maintain more connections to the outside, making them more amenable to cross-pressures and divergent cultural forms. Both core and periphery are important for the group. The core defines its boundary and identity, while the periphery provides irritations and adaptation to the sociocultural environment.

Together, these arguments translate into the analytical schema presented in section 2.1, but in a meso-version (figure 3.3). *Foci of activity* are the most important opportunity structures for groups, allowing and making for repeated contact among its members. Frequently, this entails the simultaneous activation of multiple ties. Foci of activity are one source for the dense *patterns of communication* among group members. In turn, they have to be established as meeting points in these patterns; otherwise, they lose their function for the group. Among members, communication has to be mostly collaborative, sociable, and solidary. But groupist communication patterns do not only run between members. Between a group and its environment, personal communication has to thin out, or turn conflictual, particularly between rival groups or to an environment defined as hostile.

As always in network research, such patterns of communication are guided by *relational expectations* between the actors involved—their social relationships. These include positive ties (with expectations of collaboration and solidarity) within the group, as well as negative ties of ignorance or conflict to the outside world. The relational expectations governing the course of communication here adopt prescriptions from the *group identity* and *group boundary*. In relation to the networks of personal relationships, the identity of the group and its boundary constitute a cultural environment (Bail 2014). They are institutionalized within the group, even if they lack the abstraction and generality of wider cultural patterns. As elaborated in the next chapter,

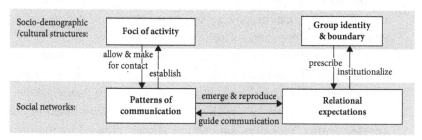

Figure 3.3 Interplay of levels in group phenomena.

group boundaries may become institutionalized across social contexts as a wider social category.

Figure 3.3. differs from figure 2.1 chiefly in the upward arrows. In figure 2.1, social categories and opportunity structure exert a one-way effect on the network patterns of communication and its meaning structure. This was overly simplistic, but probably valid for many large-scale opportunity structures and cultural models. In the case of groups, foci of activity, on the one hand, and the identities and boundaries of groups, on the other hand, have to be rooted in communication patterns and in relational expectations, respectively. Otherwise, they cease to work as meeting points or they lose their ability to convince and to mobilize. The sociodemographic and cultural structures behind group phenomena are more firmly tied to network patterns than those in the wider society—be they residential segregation or social categories like ethnicity and gender (see chapters 4 and 6).

The tight interplay of social boundary, cohesive network, and group culture seems to insulate the social group, in particular if the group manages to organize around a focus of activity. In this chapter, I focus on how group constellations are able to persist, thence the emphasis on stabilizing factors, and on the inner life of groups. However, the outside world plays an important role in four regards:

- Foci of activity are based on availability and on *spatial conditions*. But also, formal organization can facilitate or inhibit group formation. Schools and workplaces make small populations meet repeatedly. Bars and shops provide opportunities for contact among the like-minded. If more people meet at these foci, multiple groups are likely to emerge who distinguish among themselves by way of symbolic boundaries and sub-styles.

- The *social boundary* can become *politicized*, through the connection to grander social struggles. Religious practices or private behavior (like drinking alcohol) can become subject to political contestation, pitting adherents versus opponents (Gusfield [1963] 1986).

- In this, and in other ways, the outside world can provide *cultural blueprints* for group boundaries. In this way, many of the Black street gangs calling themselves "Crips" and "Bloods" in the United States lack personal or organizational connection to the famous gangs based in Los Angeles. Rather, they draw on cultural models diffused by the *mass media*. Groups also pick up on mass-mediated cultural material to develop their group style, sometimes perverse original meanings. For example, the punk subculture adopted the swastika as a symbol of their resistance to mainstream culture (Hebdige 1979, 116f). The swastika, widely known as the symbol of the totalitarian and culturally oppressive Nazi regime (that adopted it from Hinduist and Buddhist spirituality), makes for an odd couple with punk's emphasis on resistance and anarchy.

- Finally, every in-group needs an *out-group*. The group's social boundary is not only drawn to exclude nonmembers. Rather, it symbolically distinguishes the group from other, rival groups, for example, from other gangs competing for turf and symbolic status. But the "out-group" can also consist of other sociocultural structures. Deviant groups may stress their difference from the local community and from the families their members come from, from the school system, and the police. Becker's jazz musicians and marijuana smokers construct their identity against the reactions of their audiences and against dull mainstream music (1963). *Conflict*, in various versions, thus seems crucial for the persistence of groups. Maybe the outside world could effectively curtail deviant groups by ignoring, not fighting them.

Given the complex social structures of the present, and the stochastic nature of tie formation and dissolution, groups are a rare breed. With the tight interplay of networks and culture, relatively cohesive groups with homogeneity in cultural forms tend to form all the time. But they are unlikely to stabilize and persist. To persist, groups have to organize around a focus of activity (conjecture 3.1), and they have to construct a symbolic boundary through conflict with other, rival groups or with a hostile social environment (conjectures 3.3 and 3.9). In both regards, the group's environment is vital.

(b) Qualifications

Finally, a few qualifications are in order. What phenomena exactly are covered by the model? Obviously, "groups," as conceptualized here, would have to feature a degree of social cohesion, a relative homogeneity in cultural forms, and some sort of symbolic dividing line between itself and its environment, preferably a name for the group, and other symbols attached to it (a group style).

Groups differ from *formal organizations* like companies, administrative units, or universities (Kühl 2020). Formal organizations ensure their persistence with material rewards, usually an income, and with formal hierarchies that allow them to make corporate decisions even against the will of their members. A gang member might similarly mind the task allocated to him by his leader, for example, killing the member of another gang. But his motivation to comply has to come from a desire to belong to the group, and from the status conferred to him if he completes the homicide. He might even agree that it is necessary to fulfill the task for the glory of the gang, even if he does not want to kill.

Groups are ideal types contrasting with companies and universities in their informal organization, with personal ties important for recruitment and commitment, and in the emphasis on collective identity. However, the difference is gradual. Companies and universities still stress their "corporate identity." And employees are often enough recruited through social ties, which also enhance their commitment and compliance. Similarly, they frequently develop an organizational culture, with attitudes and values, symbols and jargon. Between formal organizations and groups, we find a number of hybrid cases. Social movement organizations and voluntary associations rarely provide material benefits, relying on their members' identification and commitment. This precludes treating their members only formally, rather than as persons with particular interests and ideas. But they feature formal roles like chairwoman or treasurer, and they clearly distinguish between members and nonmembers. Political parties lean further to the pole of formal organization, since they need to implement formal decision procedures that impede flexibility. Robert Michels finds parties to be dominated by their elites, rather than following the will of their members ([1911] 2001). However, like social movements and other group phenomena, they rely on a strong collective identity and a symbolic boundary between them and competing political forces. Also, they build on, and foster, a particular culture of symbols and worldviews.

I have argued that *families* do not conform to the model, due to their focus on individual persons (rather than exchangeable members). *Subcultures* of lifestyle or political conviction, *ethnic and religious groups* show many similarities with groups (Fine and Kleinman 1979; Hebdige 1979). However, they are too big to have many of their members know each other, and to organize around one or a few foci of activity. We can think of them as "networks of groups" (see chapter 4). But what do the ties between these groups consist of?

According to Fine and Kleinman, subcultures are integrated through "weak ties," through common media (journals, web forums, etc.), and through multiple memberships of individuals in groups (1979, 10ff).

Sometimes, however, the unity of an ethnic group, or a subculture, only consists symbolically, as Brubaker suggests (2004). Political elites and the mass media invoke and diffuse social categories and cultural models, and these get picked up "on the ground" in face-to-face interaction in villages, on street corners, in schools, and at the workplace. Many Bloods and Crips are aligned with their West Coast brothers only in name and symbols (Monti 1994). Similarly, punks everywhere pick up on the music and symbols of a few musicians in London and Manchester (Crossley 2015a). I would not classify these larger social structures as "groups" without the necessary social structural cohesion (definition 3.1). We need more elaborate models to account for subcultures and social categories.

This leaves us with only a very few cases that perfectly fit the ideal type of group, and the theory offered here: baseball teams, urban communes, other millenarian groups, local street gangs and communities, and the original punk scenes. Other punks, and many wannabe gangs, still form groups, but on the basis of culturally available models. The value of the model lies less in its applicability to a wide variety of perfect cases, and more in its ideal-typical depiction of mechanisms at work in the construction of group phenomena.

A final instance to consider are "invisible colleges" in academia (Crane 1972; Griffith and Mullins 1972). From the Royal Society in London, through the Vienna Circle to Los Angeles–based ethnomethodology (Mullins 1973), they have been important drivers of scientific development. Invisible colleges share the characteristics of groups: most of them are organized around particular departments or cities as *foci of activity*, particularly in their early stages; they have a densely connected *core*, and a loose *periphery* of followers; they develop a particular academic/scientific approach as shared culture of the group; and they draw a symbolic boundary around themselves, establishing their identity as opposed to competing approaches and to the mainstream. Relational sociology itself, with its early geographical center in New York City and the close personal connections among its adherents, fits the model (Mische 2011; Fuhse 2015b, 21ff).

The example of invisible colleges shows that group phenomena are not confined to interpersonal networks of friends. They appear in social fields like academia and science, in the arts (White 1993; Crossley 2015a), and probably also in other areas (political and religious movements, informal economic circles, etc.). We would have to examine their conditions and consequences in detail, and in conjunction with the forces in these fields (Collins 1998). At this point, I can only suggest that mechanisms of group formation and stabilization play out in a number of social fields, and that group closure and conflict have an impact on their dynamics.

4

Ethnic Categories and Cultural Differences

The central concern of relational sociology, as pursued by and around Harrison White, is the interplay of social networks on the one hand, and meaning and culture on the other hand (Pachucki and Breiger 2010; Mische 2011; Fuhse 2015b; McLean 2017; see chapter 2).[1] In this light, its relative lack of attention to intercultural and interethnic constellations comes as a surprise. Notable exceptions are Charles Tilly's account of social inequality as rooted in categories inscribed in networks (1998) and Tammy Ann Smith's study of the construction of ethnic boundaries (2007). Mustafa Emirbayer and Matthew Desmond offer a theory of the American "racial order" inspired by relational sociology (2015). However, building on Bourdieu, Durkheim, and Dewey, they accord more importance to symbolic demarcations, objective positions, and collective emotions than to networks of relationships. Andreas Wimmer's work on ethnic boundaries (2013), in contrast, sees ethnic categories as partly rooted in network constellations, closer to the relational sociological project. In this chapter, I draw on their work, on relational sociology in general, and on the framework from chapter 2 to develop a theory of the interplay of ethnic categories, cultural differences, and social networks.

Such a theory has to link two layers of the social world: the relational layer of relationships and networks, and the symbolic layer of cultural differences, ethnic categories, and symbols of demarcation. Previous attempts at systematizing interethnic relations have rarely embraced this duality. This chapter simultaneously applies and extends the theory of groups as symbolically bounded network segments from chapter 3. The basic interplay of network and meaning holds for both group phenomena and for ethnic categories. Ethnic categories form part of the cultural patterns adopted and inscribed in the meaning structure of social networks, as devised in section 2.1. However,

---

[1] This chapter builds on my article "Embedding the Stranger: Ethnic Categories and Cultural Differences in Social Networks," published in the *Journal of Intercultural Studies* 33 (2012). I thank Taylor and Francis for the permission to reuse the material here.

Social Networks of Meaning and Communication. Jan Fuhse, Oxford University Press. © Oxford University Press 2022. DOI: 10.1093/oso/9780190275433.003.0004

ethnic categories are wider in scope than any group identity. They rest on the cultural construction of differences in politics and the mass media, as well as in face-to-face interaction.

American sociology has frequently focused on "race" as a physical trait that acquires a certain significance in society (Bonilla-Silva 1997; Emirbayer and Desmond 2015). I follow scholars that de-essentialize race and focus on the construction of ethnic categories instead (Loveman 1999; Wimmer 2015). Ethnic groups are artificial constructs, based on a belief in common ancestry, which fosters solidary in-group relationships (Weber [1922] 1978, 389ff). This categorization is made inside the group (self-ascription), but also by other groups and in wider discourse (other-ascription; Barth 1969). Race can be seen as a subclass of ethnic categories with an emphasis on phenotypes. The mechanisms at play do not differ much—therefore, no need to deal with race in particular, rather than with the more general phenomenon of ethnic categories.

I start by picking up selectively on previous works on interethnic relations. First come the powerful figures of marginality: Simmel's and Schütz's *stranger*, Robert Park's *marginal man*, and Norbert Elias and John Scotson's *established* and *outsiders* (section 4.1). The second section deals with ethnic boundaries in anthropology (section 4.2). Then I discuss the recent social capital approach (section 4.3). I turn to the relational sociological perspective with a short summary (section 4.4). The following sections examine particular aspects of ethnic categories: their interplay with networks (section 4.5), the relational nature of ethnic boundaries (section 4.6), and network theoretical reconstructions of the marginal constellations from the first section (section 4.7). I conclude by situating this network theory of interethnic relations in the context of other approaches to ethnic groups (section 4.8).

4.1. Marginal figures

This section offers a brief overview of the early figures of marginality in the sociology of migration: the stranger (Simmel, Schütz), the marginal man (Park), and the established and outsiders (Elias and Scotson). These figures illuminate the social constellations resulting from migration, and the processes of meaning connected to it.

The classical treatments on the *stranger* by Georg Simmel and Alfred Schütz focus on the cultural side of interethnic relations. As someone who "comes and stays," the stranger becomes part of the group and relates to it socially (Simmel [1908] 1971). However, due to her different cultural imprint, she lacks knowledge of the typical routines and scripts of action necessary to make sense of the behavior of others in the new context, and to adequately react to it (Schütz [1944] 1971, 96ff). The stranger, acting in unforeseen ways and calling the common sense of the group into question, comes to be classified as not fitting, as "strange." The social group thus saves its cultural worldview from irritations, whereas the mindset of the stranger is transformed, at least gradually over time (Schütz [1944] 1971, 105). Schütz and Simmel both discuss the stranger primarily as a cultural phenomenon, as somebody who differs in worldviews and challenges the common sense of the group. The group reacts by forming stereotypes about strangers. But we learn little about the social constellations—the interethnic relations—resulting from this interchange.

Robert Park's formulations about the *marginal man* pick up on Simmel's stranger, adopting a social psychological perspective (1950, 351, 354f, 374). Since marginality is not confined to males, I refer to the marginal *person* here. According to Park, the social situation of the marginal person (a stranger in the sense of Simmel, often of "mixed-blood") leads her to combine cultural patterns from her home society with those from her receiving context. The marginal person is "a cultural hybrid, [. . .] living and sharing intimately in the cultural life and traditions of two distinct peoples" (Park 1950, 354f). This situation leads to a distinct "personality type" characterized by "the wider horizon, the keener intelligence, the more detached and rational viewpoint," but also by "emotional responsivity" and a certain "superficiality" (Park 1950, 135, 318, 376).

Unfortunately, Park's concept of the "marginal man" remains unrelated to the wider theoretical approach of the Chicago School, in particular to Park's own theory of assimilation. Herbert Blumer (1969) presents the basic ideas of symbolic interactionism, the theoretical stance of the Chicago School, as follows: Social life is characterized by the emergence and the negotiation of meaning (symbols) in the interaction in densely connected primary groups, and individuals by and large adopt the schemes of interpretation resulting from this process (see section 3.1.b). Particular situations are assessed on the basis of these interpretive schemes, and individuals (and groups) act upon this "definition of the situation." William Thomas ([1927] 1966, 166f) argues that races, nationalities, and communities develop their particular ways of defining situations.

Park and Milton Gordon apply these arguments to the situation of migrants in their theories of assimilation: whether or not migrants successfully acculturate into the host society chiefly depends on their incorporation into networks of autochthones (Park and Burgess [1921] 1969, 736f; Gordon 1964, 81). The more that migrants enter into the primary groups of the receiving context (social assimilation), the more they adopt the prevalent definitions of the situation (acculturation). Other steps in the process of integration—the lessening of ethnic division and the attainment of status equality by migrants—are expected to follow. Accordingly, successful integration of migrant groups depends on the formation of crosscutting relationships. Ethnicity and ethnic division are not maintained if they are not matched by the organization of personal relations.

Norbert Elias's *configurational sociology* similarly takes social relationships as its starting point. Configurations are constellations of individuals who are not acting individually, but always as part of the configuration ([1970] 2012). According to Elias, these constellations rather than individual capacities or resources are responsible for endowing kings with power, for the drift toward the Cold War, for the development of individual creativity and character, and for long-term changes in civilizations.

In their seminal study *The Established and the Outsiders*, Elias and John Scotson describe the social and symbolic alignment of the long-resident inhabitants of an English suburb against the newly arrived "outsiders" ([1965] 1994). Elias and Scotson identify internal cohesion as the main difference between the two groups: Due to their knowing each other for a long time and having formed personal relationships among themselves, the established residents develop a favorable collective identity and impose a deprecatory image on the recent arrivals ([1965] 1994, 150ff). These lack internal cohesion and are unable to develop a positive collective identity of their own. In line with relational sociology, Elias and Scotson link characteristics of network structure (structural separation and differential density) to the cultural forms (the boundary and the identities of the two collectivities) embedded and reproduced in these networks.

All three approaches—Simmel's and Schütz's stranger, Park's marginal person, and Elias and Scotson's established–outsider configurations—deal with the meaning associated with marginality. They point to the cultural side of ethnic categories, and to the social psychology connected to these situations. Classical assimilation theory of Park and Burgess, and Gordon, takes social relationships in primary groups as key for the integration of migrants but does not dwell on ethnic boundaries and identification much. Elias's configurational sociology lies closest to relational sociology in emphasizing the constellation of social relationships. The following approaches show a stronger concern for networks and their interplay with ethnic categories.

4.2. Anthropology

British social anthropology not only developed the network concept (see section 3.2.a); it also used the concept a number of times to examine inter-ethnic relations. Anthropology long suffered from a fissure between social anthropology (dealing with the relational side of social structures) and cultural anthropology (mainly concerned with symbolic forms). In the 1970s, J. Clyde Mitchell argued that ethnicity and ethnic behavior should be studied with regard to both the (network) structure of relations and the symbols and justifications (culture) connected to them (1974, 15f, 20ff).

Philip and Iona Mayer's study of Xhosa migrants from rural areas to East London in South Africa (1961) provides an early example of anthropological research following this direction. Mayer and Mayer found "country-rooted" or "Red" migrants to maintain close ties to Xhosa villages, to dress in traditional Xhosa style, and to uphold identification with their homeland and tribes. "Town-rooted" or "School" migrants, in contrast, immerse themselves in city life, they do not dress traditionally, and they partly leave their ethnic identification behind. While Red migrants are enmeshed in a close-knit network of other Xhosa (migrants and non-migrants), School migrants display more loose-knit personal networks with heterogeneous contacts in the city (Mayer and Mayer 1961, 288ff). The migrant lifestyles and the salience of ethnicity correspond not only with the composition (Xhosa/non-Xhosa) but also with the structure (close-knit/loose) of personal networks.

Fredrik Barth's famous account of ethnic boundaries (1969) focuses on the relations between ethnic groups. Barth does not conceive of an ethnic group with regard to its shared culture ("cultural stuff"; 1969, 15). This shared culture is a secondary phenomenon and results from the symbolic division of the group from other groups. This boundary exists in the ascription of ethnic identity and difference within the group and by other groups, and it effectively structures the interaction between them. The ethnic boundary does not fully separate ethnic groups. Rather, it ensures that interaction within groups differs systematically from that between them. Within an ethnic group, interaction is amicable and trusting. Between ethnic groups, interaction is often imprinted with stereotypes and status distinctions. Consequently, we expect different types of tie within and across the ethnic boundary: friendships, trust, and solidarity among co-ethnics on the one hand, and routinized and more formal interethnic interaction fraught with distrust and stereotypes on the other hand. According to Barth, this organization of interaction along the ethnic boundary is itself a condition for the persistence of the boundary (1969, 18).

The work of Andreas Wimmer, bridging anthropology and sociology, follows up on Barth's formulations on ethnic boundaries. These rest partly on the "social closure" of social networks within a category. Social closure makes for material inequality because social networks give access to resources (following the social capital literature, discussed further in this chapter; Wimmer 2013, 83ff). In contrast to Barth, Wimmer argues that an ethnic group's cultural stuff can make an important difference: cultural difference between two groups reinforces the ethnic boundary, just as a salient boundary enhances cultural difference (2013, 86f).

In his empirical study of migrants in Switzerland, Wimmer finds that the Swiss, the Turkish migrants, and the Italian migrants form friendships and other close relationships largely among themselves (2013, 126ff). All of these groups apply "classificatory schemes" that allow them to view their own group as valuable, and to denigrate others (2013, 118ff). This mirrors the closure of the network of established residents in Elias and Scotson's study through the advantageous collective identity.

Wimmer's wider discussion of the "making" of ethnic boundaries points to their variability (2013, 49ff): they can be expanded and amalgamated (e.g., in nation building), the ethnicity of minorities can emerge ("ethnogenesis") in contrast to that of a national majority, they can be crossed by individuals or shift in the evaluation and social status associated with them, or they can blur altogether. Networks of personal relationships probably play a role in these processes. However, Wimmer mainly discusses political mobilization, public discourse, and institutions as more important vehicles of ethnic boundary making (2013, 63ff).

4.3. Social capital

Sociology often discusses the relevance of social networks for questions of social inequality in terms of "social capital." Much of the recent literature on social networks of migrants picks up on this somewhat fuzzy concept. To speak of social capital, as proposed by Pierre Bourdieu, James Coleman, Ronald Burt, and Nan Lin, implies that social networks afford advantages to the actors embedded in them (see section 7.3). But important differences prevail:

- For Bourdieu, social capital consists of the resources that can be activated through friends or acquaintances (1986).
- Coleman finds social capital to lie in the propensity of close-knit networks to enforce norms and ensure cooperation among its members (1988, 1990a, 1990b).
- Burt stresses the importance of "weak ties" (picking up on Granovetter; discussed further in this chapter) across structural holes in network structure for access to information (1992).
- Lin combines Coleman's and Burt's approaches: Strong ties in dense networks should be important for expressive needs ("bonding"). Weak ties across structural holes help for instrumental action, due to providing access to information from distant network regions in social structure ("bridging"; 2001).

All four authors argue that networks are "good" for actors. But the aspects of networks considered "good" differ profoundly.

We find the earliest link to interethnic relations in Mark Granovetter's article on "The Strength of Weak Ties" (1973), one of the inspirations for the social capital approach. Granovetter argues for the importance of weak, somewhat superficial contacts in order to obtain important information (that is not transmitted in the network of close friends and family) or to organize on a community level. In this latter context, he tentatively explains the lack of community organization among Italian migrants in Boston's West End (in an earlier study by Herbert Gans) with the absence of bridges (weak ties) between the families and cliques (1973, 1373ff). This points to the importance of network structure in migrant communities.

In their 1993 study, Alejandro Portes and Julia Sensenbrenner introduce the concept of social capital into the sociology of migration. They take two important steps to a differentiated notion of social capital: First, they disentangle the effects of networks on migrants by identifying four mechanisms: the introjection of values in networks, the reciprocity of transactions, the emergence of bounded solidarity, and of enforceable trust in densely connected migrant groups (1993, 1323ff). Second, they demonstrate that the same mechanisms can help or constrain migrants, depending on circumstances (1993, 1338ff). For example, high degrees of connectivity in migrant communities can be detrimental if these communities emphasize solidarity and connection to the old home rather than upward mobility in the receiving context. Portes terms this the "dark side" of social capital (1998, 15ff). He and Sensenbrenner only focus on dense social networks ("closure" in the sense of Coleman) rather than weak ties (Granovetter/Burt) as social capital.

Other recent studies similarly point to the importance of networks in multiethnic contexts. Susan Hardwick (2003, 169) and Paula Pickering (2006, 80, 83ff) pick up on Lin's distinction between bonding and bridging social capital to argue that social relationships that help in coping with problems (bonding or "social support") run mostly within ethnic groups. In contrast, bridging social capital that fosters upward mobility ("social capital as leverage") consists mostly of interethnic ties.

Overall, the social capital concept may be *en vogue* and sound convincing, but it does not fully clarify the role of networks in migration. Following Portes and Sensenbrenner, we have to acknowledge that networks can have very different effects according to network structure and cultural context. Also, the social capital approach focuses on network effects and remains mostly silent on why networks are structured in a particular way. It seems more promising to disentangle various network mechanisms than to broadly subsume them under the umbrella notion of social capital.

4.4. Interlude

This overview is far from exhaustive. But it identifies a number of core issues that any account of interethnic relations has to deal with. All the approaches agree that networks of personal relationships play an important role in interethnic contact. Some only deal with the composition of social networks—whether there are many or few interethnic relationships (social assimilation). Others point to the structure of networks: Elias and Scotson, and Granovetter, argue that ethnic minorities lack internal cohesion. In Mayer and Mayer's study, the upwardly mobile (School) migrants had looser networks (more weak ties) than the more traditional Red migrants.

As the Chicago School, anthropology (Mitchell and Wimmer), and Elias argue, networks link to culture and meaning. Park and Gordon view migrant acculturation as resulting from social assimilation. Following Barth, ethnic boundaries depend on the organization of networks along ethnic lines, but they also foster this structural division between groups. Contra Barth, Wimmer argues that the cultural stuff of ethnic groups can enhance this structural separation. The contact between ethnic groups is effectively organized by classificatory schemes that ethnic groups develop.

Following these impulses, I examine cultural differences and ethnic boundaries in conjunction with social networks. In the following pages, I sketch a network theoretical approach to interethnic relations building on the general framework from chapter 2 and on the theory of group cultures and boundaries from chapter 3. In this perspective, social networks are not only the basis for the emergence and negotiation of culture but are themselves symbolic in nature. Cultural differences and ethnic categories should be closely connected to social networks.

4.5. Ethnic categories and networks

Categories like class, race, or gender are central to the constitution of social inequality. In empirical social research, such categories are often treated as "independent" or as "control variables" as if they were given and directly causing social inequalities (Martin and Yeung 2003). Relational sociology, in contrast, views social structures as resulting from the interplay of categories (as symbolic forms that govern transactions) and networks. Whether or not categories bear an imprint on social inequality depends on patterns of social relationships and on the meanings attached to the category (Tilly 1998). I have argued in chapter 2 that social relationships are not all alike, and that we should not just treat them as "on" or "off." Instead, relationships entail particular communication patterns and relational expectations. They come in various strengths, and with different relationship frames (each with its distinct set of expectations). In the following, I first write about social relationships only as strong, positive ties, that is, friendships, family ties, and love relationships, here summarized as "personal relationships." This is in line with the theoretical formulations of White, Gould, Tilly, and Wimmer. Later, I return to a more multifaceted perspective, with a concern for ties that are weak or conflictual.

In an influential, but long unpublished paper, Harrison White argues for "catnets" (networks that are structured by categories) as one of the building blocks of social structure ([1965] 2008, 8ff). The three characteristics of catnets are: (1) a social category separating inside from outside, (2) increased density of ties inside and few ties to the outside, and (3) similarity of cultural forms within the network—a particular style that distinguishes the catnet from other catnets. A catnet either emerges because a category increases the likelihood of internal ties, which in turn leads to similarity of cultural forms. Or the catnet results from the categorization of a network with its particular cultural imprint (the domain) as different from other networks.

These formulations remind us of the theory of groups as network segments stabilized by a symbolic boundary from chapter 3. However, categories are larger in scope and more institutionalized than the boundaries of singular groups, forming part of the wider cultural repertoire that structures interaction and relationships. White's arguments about categories are compatible with group identities that emerge endogenously, out of the internal interaction in a densely knit network. However, my concern here is with large-scale ethnic categories that span many network contexts. These certainly rely on face-to-face networks patterned accordingly. But they are also constructed and diffused in politics, education, and mass media as large-scale infrastructures of communication and symbolic construction (see section 2.2). And they display a certain generalization because they apply to wider categories of people rather than to concrete circles of interaction. Given the strong similarity of ethnic categories and group boundaries, we revisit many of the mechanisms from chapter 3.

Charles Tilly picks up on White's concept of catnets in his theory of inequality (1998). For Tilly, the closure of networks by categories is an important mechanism explaining durable inequality. Social categories are symbolic forms that lump together and split sets of actors by stressing their similarity or dissimilarity and "define relations between the two sets" (Tilly 1998, 62ff). The categorical division is accomplished in the form of "stories" arising and circulating in transactions, but also constraining subsequent transactions. Through these stories, categories effectively structure social relations, fostering trust on the inside and mistrust to others (Tilly 2005b, 43f). Groups of people sharing the same category can "hoard opportunities," barring others from access to precious resources (Tilly 1998, 91ff, 153ff).

> Conjecture 4.1. Ethnic categories are rendered plausible through narratives about the ethnic differences at play.

Tilly's formulations echo the arguments by Portes and Sensenbrenner about the reciprocity of transactions, the emergence of bounded solidarity and of enforceable trust in migrant groups, and their disentangling of network effects into various mechanisms. Unlike Portes, Tilly shies away from the fashionable term "social capital," probably to avoid the confusion and the pitfalls connected to it. The ethnic category's imprint on transactions reminds us of Barth's theory of ethnic boundaries. Tilly also draws on core ideas of relational sociology: identities are defined and related to other identities in stories (Somers 1994; Tilly 2002, 10ff), and social entities only exist because of the boundary separating them symbolically from their environment (Abbott 1995).

In contrast to most empirical social research, Tilly does not take the category as given or as independent from the transactions in networks. Rather, the salience of a category results from convincing storytelling, from its being taken up in transactions, and the resulting effective structuring of social relations. As Roger Gould argues:

> The social categories (class, race, nation, and so on) within which individuals see themselves as aligned with or against other individuals depend on the conceptual mapping of the social relations in which they are involved and on the partitioning of people into collectivities whose boundaries are logically implied by this mapping. (1995, 17)

Categories rest on social relationships being structured along their lines. If network structure changes, so does the salience of categories. On the other hand, boundaries make for an ordering of social ties. They work as schemas of classification for social structure, as boundaries for the construction of social identities:

> Conjecture 4.2. Ethnic categories structure networks of social relationships, but they also depend on their reproduction in these. Ethnic groups with

higher structural separation in personal relationships display a higher salience of the ethnic category.

Translated into the framework from chapter 2, the ethnic category is a cultural pattern drawn upon to structure personal relationships. It offers a blueprint of "relational expectations" within and across the category. Inscribed in the meaning structure of social networks, the category also affects the communication taking place, making for ethnically organized patterns of interaction (Rytina and Morgan 1982).

The group structures in chapter 3 are organized around a common core (conjecture 3.7), and they frequently build on one or two activity foci for their members to meet (and social relationships to form; conjecture 3.1). Ethnic categories lack this centralized organization, and they need a lot more foci of activity. Accordingly, their structural separation rests on enhanced opportunities for contact among co-ethnics (Blau 1977). Of course, ingroup ties are also fostered by other mechanisms, for example, by linguistic differences. But these lose much of their importance when minorities acquire proficiency in the majority language. Therefore, the salience of ethnic distinctions at least partly depends on differential opportunities for contact. Such differences rest on the geographic concentration of ethnic categories in certain neighborhoods, and on the differential access to schools and universities, jobs, and leisure activities (Mouw and Entwisle 2006):

Conjecture 4.3. The structural separation of personal relationships by ethnic categories correlates with the opportunities for interethnic contact in schools, neighborhoods, in leisurely activities, and at the workplace.

We can thus theoretically account for the high ethnic homogeneity of personal relationships—and predict under which circumstances intraethnic homophily and the salience of ethnic categories should decline. It should be possible to break the vicious circle of ethnic categorization and network segregation by (a) changing the composition of neighborhoods, at the workplace, or in schools, or by (b) successfully challenging and reframing the stories connected to an ethnic category. Both may result from unplanned processes or from deliberate attempts by political authorities, social movements, or economic or cultural elites.

Empirical studies on interethnic personal relations have yielded extensive evidence for the tendency of ethnic groups to form intimate relationships primarily among themselves. Friendship networks are denser within ethnic groups than across racial boundaries (Louch 2000, 58). Schweizer et al. (1998) find the personal networks of Hispanics in southern California to comprise more kin than those of Whites, and to be more locally concentrated. Much of this can be explained by reference to opportunities for contact, for example, at the workplace or in schools (Fong and Isajiw 2000, 261ff; McPherson, Smith-Lovin, and Cook, 429ff; Moody 2001; Wimmer and Lewis 2010; McFarland et al. 2014). However, research on networks in schools shows a tendency to form intraethnic friendships even after controlling for classroom composition (Hallinan and Williams 1989; Quillian and Campbell 2003; Baerveldt et al. 2007; Currarini, Jackson, and Pin 2010; Smith et al. 2016).

4.6. The relational nature of ethnic boundaries

These empirical findings demonstrate the general tendency of ethnic or racial groups to form intra-category ties. What is it about ethnicity that makes people treat people from other ethnic groups differently? Following Max Weber, an ethnic category can be defined as the symbolic boundary dividing different groups along criteria of believed descent ([1922] 1978, 389; Yinger 1985). It may be coupled to religious or linguistic differences but need not be. As in Elias and Scotson's study ([1965] 1994), ethnic distinctions do not have to rest on long-standing cultural differences. It is more important that the ethnic category be used as a symbolic device that effectively structures communication and the formation of relationships (Barth 1969).

At this point, the argument turns into a tautology. How can I argue that social networks are effectively structured by ethnicity, which is defined as dividing groups in the first place? We can disentangle this conceptual mess by defining ethnic categories as schemas of perception and differential treatment of other people (bearing in mind the criteria of descent that distinguish ethnicity from other social categories like gender or age).

Definition 4.1. An ethnic category is a schema that classifies people by their (supposed) common ancestry and comes with particular expectations about how to interact within the category (among co-ethnics) and across it (in interethnic relations).

In this sense, the ethnic category is a frame people draw on to define social situations. Unlike in Thomas's formulation (see section 4.1), it is not the specific worldview of an ethnic group with particular values and other cultural patterns that features as the definition of the situation—the distinction between ethnic groups itself does (Merton 1958, 477ff). If people are classified by (believed) descent, if particular ethnos-typical behavior is expected of them, and if they are treated differently, then communication and social relationships are framed and structured along ethnic lines. Ethnic categories come with expectations for intraethnic and interethnic contacts (Barth 1969). This includes general rules for what kinds of people to develop what kinds of relational expectations to. The ethnic category is thus inscribed in the meaning structure of social networks. This holds true even for interethnic ties—an interethnic friendship or love relationship will feature the deviation from the expected norm in its "relationship culture." As long as most ties are formed according to cultural expectations, the ethnic category has an impact on the composition of networks of positive personal relationships (conjecture 4.2). Like relationship frames, ethnic categories are institutionalized models for organizing communication (see section 2.7 and chapter 5).

These formulations point to an important shift in the concepts of "frame" and "definition of the situation." These are not conceptualized as cognitive, but as part of the process of communication and inscribed in social structures. In contrast to the sociology of cognition, ethnicity is not seen primarily as cognitive (Brubaker 2004, 64ff). Like other social categories, ethnicity has to be communicated, "done," and not only thought, in order to become a lasting feature of the social world (West and Fenstermaker 1995).

This "doing" of ethnicity can lie on different levels: It can refer to classificatory schemes people use in their communication to categorize ethnic groups, as in Wimmer's research in Switzerland (see section 4.2). It is performed with the telling of stories about common ancestry and behavioral traits connected to it, as Tilly argues (see section 4.5). Ethnicity is also found in moral evaluations (Lamont 2000, 55ff, 169ff). It can consist of differential treatment of people falling into different ethnic categories (Massey 2007). Or ethnicity can lie in customs that "flag" people as ethnic (Billig 1995, 38ff)—for example, the wearing of a Jewish kippah or of a Muslim chador. In any case, the ethnic category is part of "culture in use," rather than of an abstract universe of ideas (see section 2.2.b).

Drawing on Bourdieu, observable displays of ethnicity can be termed "ethnic practices" (Bentley 1987). This practical side of ethnicity contrasts with ethnic identification, which can become "symbolic" and unconnected from the actual doing of ethnicity (Gans 1979; Alba 1990). Some of these aspects of ethnicity are other ascribed (like differential treatment); others are done within the ethnic group (self-ascription). As Barth hypothesizes, these should not be independent (1969, 13f). Self-ascription is observed in other-ascription. And discrimination by others probably leads to a turn toward one's own ethnicity.

> Conjecture 4.4. Ethnicity is multidimensional, with ethnic identification and differential treatment interplaying with ethnic practices.

An important instance of ethnic practices is the use of particular languages. Language can support ethnic division by serving both as a symbol of difference and as a medium restricting the communication between two groups (Lieberson 1981). Of course, the acquisition of language skills is important for upward mobility of disadvantaged minorities. But language also organizes the communication in networks (Mohr 1994; White 2008, xixff, 28, 31). Speaking a minority language—in school classes, at the workplace, or even in politics—marks the ethnic category as important in communication (it "does" ethnicity) in a way not easily circumvented.

> Conjecture 4.5. Networks are organized through linguistic practices. The knowledge and practice of languages is of central importance for the reproduction of structural separations in migrant networks.

In line with the arguments presented earlier, these various forms of "doing" the ethnic category—from discrimination to linguistic practices—should correlate with the composition of personal networks. It is difficult to form intimate relationships across the barriers of discrimination and language, especially if the ethnic categories involve expectations to not form intimate interethnic ties. But of course, this is a matter of degree: the salience of a category should be measurable looking at various forms of ethnic practices, and these different dimensions vary according to circumstances. For example, an ethnic group can emphasize its difference without being discriminated against. But all of these forms of practicing ethnicity should be linked with the ethnic composition of networks.

Conjecture 4.6. The composition of social networks is more closely connected to the "doing" of ethnicity in ethnic practices (such as language use and ethnic clothing) than to ethnic identification.

However, the salience of a category depends not only on social-relational practices (of recognition and discrimination) but also on public discourse, political mobilization, and even state-administered discrimination between different groups (Brubaker 2004; Wimmer 2013, 63ff, 90ff). This is the first mechanism underlying the emergence of a catnet (White), as argued previously (4.5). The category is here imposed from the outside, increasing the likelihood of forming relationships within the catnet, which then leads to cultural similarity. In line with the second mechanism, a category becomes more or less salient depending on whether it matches cultural differences in networks. Tightly knit personal networks lead to a distinct lifestyle, which, in turn, will be classified into ethnic categories in public and private discourse (Shibutani and Kwan 1965, 202ff). Now the ordering of personal networks partly depends on the salience of ethnic categories in the first place, which makes for the cycle of reproduction of ethnic lines of division: salience of categories—networks—lifestyles—salience of categories.

Conjecture 4.7. Higher internal connectivity and structural separation of personal relationships in ethnic groups lead to the development and reproduction of distinct cultural practices and values. These are classified as "different" and strengthen the salience of social categories.

Do these theoretical arguments resonate with empirical research? With the prevalent methods of survey research, it is easier to ask individuals for their subjective ethnic identification than to observe ethnic categories in communication. Miranda Lubbers and her coauthors show that African and Latin American migrants in Spain with more interethnic relations to Spanish people and to migrants from other countries identify less often ethnically with their home country (2007). Lars Leszczensky and Sebastian Pink find that ethnic homophily in school friendships in Germany is primarily driven by adolescents with high ethnic identification (2019). Ethnic categories affect social relationships only to the extent that they become inscribed in individual minds (as identification) and communication (as ethnic practices and classification). Conversely, autochthones seem to become homophilous only when facing dense groups of immigrants (Smith et al. 2016). Ethnic categories become salient in network composition with sufficient numbers of minority members to meet and form ties, and this could lead to a "closing of ranks" among the majority.

My own research on Italian migrants in Germany points to ethnic identification being more symbolic and less correlated to the composition of personal networks than ethnic practices (language use) and alienation from their home country (Fuhse 2008, 166ff). Italians in Germany are less subject to stereotypes and discrimination than most other ethnic minorities considered in the literature, and this might allow for a certain decoupling of ethnic identification and social relationships, in line with Gans's concept of symbolic ethnicity (1979; Alba 1990). Ethnicity as the boundary between ethnic groups here consists primarily of ethnic practices, such as language use, discrimination, or ethnic customs, not of ethnic identification.

Tammy Ann Smith's work on conflicts between Italians and Slavs in and from Istria is a rare study of the construction and negotiation of ethnic boundaries in storytelling (2007). Smith demonstrates that the ethnic category is used to make sense of complex historical processes and personal experiences, and that the clashes between Italians and Slavs in Istria transform into conciliation once migrants from the different groups meet in a new context (New York City). Unsurprisingly, these different kinds of storytelling correspond with network segregation in Istria and with the formation of interethnic personal relationships in New York City.

The construction of panethnic categories and their relevance in personal relationships offers an important test of the theory (Okamoto and Mora 2014). Categories like "Asian American" or "Hispanics" in the United States, or Arab migrants in Germany, encompass a number of ethnic groups from different nation-states (Korean, Chinese, etc.; Mexican, Cuban, etc.; Syrian, Egyptian, etc.). They are examples of the "amalgamation" of ethnic boundaries (Wimmer 2013, 50f, 108). These categories may arise first out of classificatory practices by the state, media discourse, and ethnic majorities. But there is a firm tendency toward "panethnic friendships" in addition to a strong preference for friends from the same ethnic group, especially where minority groups are relatively small (Kao and Joyner 2006). The panethnic category unites even groups from very different cultural backgrounds, like India and Korea, symbolically. This discussion also shows that ethnic categories are not strict either/or constructions. They vary in their salience, but they can also overlap and be nested in multiple levels of categories and subcategories of wider and smaller mesh. These together make for the complex patterns of social association and symbolic classification.

4.7. Networks of marginality

How do these arguments help us understand marginality, as captured in the figures of the stranger, the marginal person, and established–outsider configurations? The stranger and the marginal person are not discussed with regard to their embeddedness in social relationships in the essays by Simmel, Schütz, and Park. We can describe the stranger as somebody from a distant network cluster entering a new social environment (with its unique cultural framework). Here, she is categorized as alien and treated accordingly. The category emerges as a definition of the situation of cultural difference. "Strangeness" is part of the storytelling connected to the category that serves to distance the long-time residents from the outsider symbolically, and to shield their culture from irritations. Social structure, in contrast, changes profoundly: the previously homogeneous social world is riven by cultural differences and social categories.

The established–outsider configuration of Elias and Scotson is a social structure resulting from this process, as I will discuss further. Other social constellations are also possible, for example, one of two coherent groups competing against each other, or even the newcomers dominating the autochthones due to military force or other resources. Whether or not this incorporation of strangers into social structure leads to a lessening of cultural difference and ethnic division in the long run depends on the structural pattering of social networks, and on the storytelling connected to the category.

Robert Park's account of the marginal person explores the social psychological correlates of the sociocultural situation of the stranger. While the culture of the established residents protects itself with the category, the marginal person has to gradually acquire the knowledge of the receiving context in order to get along. But due to her original cultural imprint and to continuing ties to the old context, this cultural framework never quite becomes as unquestionable for her as for the autochthones.

Here, the nature of culture changes. Non-strangers are wholly enveloped in their culture, which tells them how to define situations, and how to act in them. The marginal person, in contrast, learns about different ways of confronting situations—culture becomes a "tool-kit" (Swidler 1986) from which she can choose among different available cultural patterns. Swidler refers to the different social situations that lead to these two contrasting modes of culture as "settled" and "unsettled lives" (1986, 278ff). For people living in a relatively stable environment, culture provides unquestionable models for how to frame and deal with situations. In "unsettled lives" (migration being one example), culture becomes a "tool-kit" or repertoire from which to choose different ways of defining situations, or different courses of action.

If we transpose Swidler's arguments to social networks, settled lives translate into culturally homogeneous networks of personal relationships (as we find within ethnic groups or other catnets), and unsettled lives equal heterogeneous networks (DiMaggio 1997). The marginal person has access to diverging cultural frameworks due to her interethnic ties. For people without interethnic ties, the cultural framework is not put into question. On a purely numerical level, people from majorities have fewer interethnic ties than those from minorities (Blau 1977). Their cultural framework is shielded from irritations by treating migrants as strangers. The migrants, in contrast, have to deal with contrasting frameworks and to adapt their understanding of the world.

Up to now, I have focused on categories that map and separate internally cohesive subgroups in social structure. Donald Horowitz distinguishes between two types of ethnic divisions (1985): First, ethnic groups can be largely separate, occupying different territories, and only meeting each other at the borders. Second, ethnic groups can occupy different positions in the social strata and interact frequently, but they do so in specific ways and usually with some degree of subordination. This is the classic case of most migrant groups in contrast to territorially separated groups in multiethnic states. On a network level, then, many ethnic groups are not really interaction groups based on a higher density of relationships on the inside, in the sense of White's catnets. Rather, they occupy "structurally equivalent" positions in social structure, as elaborated in White's later work (see sections 2.8.c and 5.1).

Categories do not always demarcate internally cohesive subgroups. For example, kinship categories like fathers or mothers designate particular positions in a network without increased likelihood of ties within the category (see chapter 5). The nodes of one category (e.g., "fathers") are united by their similarity of connections to other categories (mothers or daughters). White termed such categories that subsume nodes in networks with specific connection profiles "structurally equivalent" (Lorrain and White 1971). Structurally equivalent nodes are not necessarily connected to each other. Classic cases of structural equivalence are slavery or the Indian caste system, where categories (slaves/owners, different castes) mark particular positions in social structure, with specific connections to other categories. Karl Marx's allotment farmers similarly constitute a structurally equivalent class of actors who lack connections to each other ([1852] 2005, 83ff). Clients in a patron–client structure are also primarily connected to their patrons and rarely among themselves.

The established and the outsiders in the study by Elias and Scotson are a case of structural equivalence. The established long-time residents form a cohesive subgroup with a strong collective identity—a catnet. The newly arrived outsiders do not form equally dense internal networks but remain relatively isolated, similar to the allotment farmers in Marx's account. Like these, they fail to arrive at a favorable collective identity to counter the symbolic power of the established group. The established–outsider configuration constitutes a particular type of network constellation similar to that of a patronage structure or a slavery system (as sketched in figure 4.1). The slaveholders or the patrons are tied to each other and to their slaves or clients, respectively, whereas these are only sparsely connected, if at all. While the established residents form a cohesive group, the outsiders are only structurally equivalent. Following Elias and Scotson, the ethnic category results from the network structure as much as it reinforces it.

Total isolation is a rare condition in contemporary society. Migrant groups typically have a number of healthy relationships among themselves, a lot of them to family members (Schweizer et al. 1998). Following Granovetter's account of the Italian community in Boston's West End, this tendency to form ties of kin may be responsible for their lack of political mobilization (see

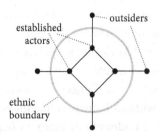

Figure 4.1. Ideal-typical network structure of an established–outsiders configuration.

section 4.3). The Italian immigrants lack the weak ties necessary to integrate the cliques and families within the community.

Such weak ties (or the lack of them in particular parts of the network) are an instance of structural equivalence. Weak ties allow reaching a faraway social position with few intermediate network steps. They afford actors with more information from different sources and with better opportunities for action and brokerage. For example, they provide access to jobs (Granovetter 1973). The difference between actors with such weak ties and those without is directly relevant for social inequality. Marginality is not only a matter of classification and social closure, but also of access to information and resources through social networks. Migrants typically have fewer of these weak ties at their avail, though interethnic relationships tend to be weak ties more often than those to co-ethnics (Louch 2000). The formation of interethnic relationships matters not only for acculturation and the lessening of ethnic categories and stereotypes, but also for the upward social mobility of disadvantaged minorities.

Granovetter's argument is mainly structural, about strong ties embedded in dense networks on the one hand, and weak ties between people without common acquaintances on the other hand. But it is also about different types of tie, in this case weak ties (e.g., between work colleagues) versus strong ties of friends and family. Therefore, we have to revise the earlier arguments. The increased density of networks among co-ethnics holds only for strong personal relationships like friendship, love, or kinship ties (conjectures 4.2, 4.3, 4.5–4.7). Cross-ethnic ties are most often weak ties, for example, among colleagues or neighbors (where workplace and neighborhood bring different ethnic groups in contact). Looking at the content of ties, it is more likely that former or current colleagues can help with job mobility than family members—provided one does not work in family-owned company, as is frequently seen in ethnic niche economies.

In Horowitz's second type of stratified ethnic divisions, cross-ethnic contacts are expected to be asymmetric, with members of the lower group deferring to the dominant ethnic group. And the ethnic boundary supposes intraethnic ties to be solidary. It symbolically pits different ethnic groups against each other—leading to conflictual ties. In line with Barth's thinking about ethnic boundaries, these do not just increase ties within ethnic groups and forbid cross-ethnic ones. Rather, relationships among co-ethnics are expected to differ in kind (friendship, love, kinship, solidarity) from interethnic ties (weak ties, conflict). Ethnic categories are inscribed in a complex web of relationships (of different types), rather than making for simple separation in networks.

> Conjecture 4.8. Ethnic categories structure the contact across the boundary, specifying how members of different groups can and should relate. This leads to situations of structural equivalence (across different types of ties), for example, in established–outsider configurations.

4.8. Conclusion

The empirical research surveyed here and the conceptual advances of relational sociology point in a similar direction: personal relationships and networks are of central importance in the construction and persistence of ethnic boundaries. This runs counter to approaches focusing on individual resources (education, income, language proficiency) as determinants of the integration of migrants (e.g., Esser 2004). While important, these resources have to be studied in conjunction with interethnic relations and with the symbolic side of migration: cultural differences and ethnic boundaries.

While I stress the importance of ethnic categories, I do not see them as chiefly responsible for inequalities, as Tilly, Michèle Lamont (Lamont and Fournier 1992; Lamont and Molnár 2002, 171ff), and Douglas Massey argue (2007). Rather, ethnic categories build on the classification of actual cultural differences and on the separation of networks of personal relationships. Ethnicity is not only constructed in public institutions (e.g., citizenship) or in political discourse, as Rogers Brubaker suggests (2004). Ethnic categories both map and structure disparities in social structure, and they are tied to social networks, institutions, and cultural differences, as well as to socioeconomic inequalities (Wimmer 2013).

With this macro-context, ethnic categories differ from the group boundaries examined in chapter 3. Group identities and boundaries always emerge and reproduce in interaction—both within the group and with its environment. As a result, they are tightly connected to the distinct cultural repertoires fostered in the dense interaction within the group. Often, they become "styles" through which groups distinguish themselves from outgroups, or from the outside world in general. Ethnic categories, in contrast, have to be proclaimed in wider discourse. A number of people interacting with each other cannot just invent themselves as an ethnic group or claim that some newly arrived residents form an ethnic group. Rather, such narratives have to survive in the pool of social classifications in political and mass-mediated discourse (Bail 2012, 2014). Ethnic categories have a higher degree of *institutionalization* than group boundaries, since they govern disparate network contexts (see chapter 5).

Ethnic categories resemble group boundaries in how they work: They provide rules for the interaction within and across the categories. They foster strong positive ties like love, friendship, or kinship between members of the same category. And they make interethnic contacts more impersonal. As a result, networks of personal relationships show increased density, as in White's model of a catnet. However, personal ties are not only patterned by ethnic categories; they also follow opportunities for contact in institutions of education, in the neighborhood, or at the workplace. This same mechanism holds for groups. They need meeting points, or foci of activity, to gather and to maintain the dense interaction among its members. But ethnic categories are larger in scope. Therefore, they depend on the widespread segregation of opportunities for contact to structure personal ties (Blau 1977).

In addition, personal relationships are more likely to form among people with similar attitudes, tastes, and lifestyles (see section 2.8). Consequently, social networks are more or less ethnically homogeneous, depending on social, cultural, and geographical contexts. If, and to the extent that, interethnic personal relationships become more common, the ethnic category loses persuasion. It becomes harder to classify people as different on the basis of their ethnic descent if they mingle more to find and develop similarities among them. As a result, the ethnic category may depend more on mapping social networks than on being able to effectively structure them. In any case, ethnic categories vary in their salience, rather than operate in an either/or fashion.

The stranger, the marginal person, and established–outsider relations can be embedded in this conceptual framework. The stranger is someone who comes into a culturally different network context and is framed and treated as alien by the autochthones to shield their way of life from irritation. The formation of interethnic ties allows the stranger to acquire proficiency in the cultural framework of the receiving context, thus becoming a marginal person. Such access to two (or more) cultural contexts leads to culture becoming a tool-kit. As in the case of established–outsider relations in the study of Elias and Scotson, ethnic categories need not correspond to preexisting cultural differences. Here, the symbolic boundary barred a heterogeneous group of newcomers from the lifeworld and the resources of the established inhabitants. These were able to impose their definition of the situation on the outsiders by virtue of their dense internal connections, building on and stabilizing a constellation of structural equivalence.

5

Roles and Institutions

Roles and institutions have been central sociological concepts since the middle of the 20th century. Much like groups, roles had their conceptual heyday in the 1950s and 1960s. They promised to bridge processes of interaction with large-scale structures of society, and to mediate between social determination and agentic creativity. But as interactionism, action theory, systems theory, and empirical research drifted apart from 1960 onward, the role concept, located between these strands, dropped out of focus. It still features in introductory courses and textbooks, and in one particular subfield: the sociology of organizations. But roles are rarely discussed by contemporary theorists. Institutions fared better than roles, even though they occupy a similar space between paradigms. Picking up on early formulations in German sociology, Peter Berger and Thomas Luckmann propagated institutions internationally in their seminal book *The Social Construction of Reality* ([1966] 1991). The neo-institutionalists around John Meyer and Paul DiMaggio adopted the concept around 1980. Ever since, institutions feature prominently in sociological discourse, but with sometimes wildly varying connotations.

This chapter assumes that both concepts (roles and institutions) map important aspects of the social world. In contrast to individualist or holistic thinking, they are inherently relational: they pertain to social relationships between actors, and to the patterns of these relationships. Little wonder that network analytical methods like blockmodel analysis aim at discerning roles and institutions (White, Boorman, and Breiger 1976; DiMaggio 1986). However, we need a thorough theoretical understanding of their interplay with social networks. I argue that roles and institutions make for a *symbolic ordering of social networks*. In the framework from chapter 2, institutions form part of the cultural patterns that are inscribed in the meaning structure of social networks. Here they become roles—typical patterns of expectations between network positions. Thus, I reformulate both concepts in the context of relational sociology and connect them to its key problem: the interplay between network structure and meaning.

Social Networks of Meaning and Communication. Jan Fuhse, Oxford University Press. © Oxford University Press 2022.
DOI: 10.1093/oso/9780190275433.003.0005

Unlike the group boundaries and ethnic categories from chapters 3 and 4, roles and institutions rarely delineate cohesive network segments. Rather, they mark relations of *structural equivalence* (Lorrain and White 1971). Actors play the same role in a network context if they resemble each other in their connections to others. These roles, and the relations between them, are institutionalized to varying degrees. That is, they follow cultural models. Social categories and relationship frames are examples of such "relational institutions." They form part of the cultural repertoire that communication draws on to construct social relationships. Hence, relational institutions make for the adoption of roles in networks, and for the systematic relationships between the occupants of these roles. But communication can always deviate from prescribed models, making for new ways of relating. Potentially, these can institutionalize and make for change in cultural repertoires and in ways of relating.

I first summarize the discussion of roles in network research, in particular in blockmodel analysis (section 5.1). Then, various strands of sociological theory are examined with regard to key concepts and arguments for the relational character of roles and institutions: Ralph Linton, Ralf Dahrendorf, and others conceptualize roles as bundles of normative expectations governing the relations between social positions (section 5.2). Symbolic interactionists Erving Goffman and Ralph Turner stress the creative deployment and modification of roles in interaction (section 5.3). The German tradition of philosophical anthropology argues that roles and institutions are cultural tools that human actors use to facilitate social life (section 5.4). Peter Berger and Thomas Luckmann, Talcott Parsons, and Niklas Luhmann shed light on the institutionalization and implementation of cultural patterns (section 5.5). Neo-institutionalism sees modern actorhood itself as a culturally variable institution for making sense of, and constructing, the social world (Meyer and Jepperson), and it suggests that we can reconstruct the institutionalized imprint of social relations in a field by way of network analytical techniques (DiMaggio; section 5.6). Relational sociologists Harrison White and John Mohr propose that roles and institutions are part of the general interplay of network patterns and meaning (section 5.7). Next, I integrate the arguments discussed into the overall perspective of this book: that social networks are bundles of relational expectations imprinted with cultural forms, developing over the course of communication and structuring it in turn (section 5.8). Finally, the social constellations and symbolic forms discussed in previous chapters are related to the two concepts of roles and institutions: group identities, collective actors, social categories, and relationship frames (section 5.9).

5.1. Network research

Roles have been more prominent in network research than institutions (McLean 2017, 90ff). Building on the classic treatment by Austro-Australian anthropologist Siegfried Nadel (1957), networks have been studied with regard to roles like "brokers" or "hangers-on." Nadel defines social structure (citing Talcott Parsons) as "the pattern or network (or 'system') of relationships 'between actors in their capacity of playing roles relative to each other'" (1957, 12). Note the difference to the earlier programmatic statement of Nadel's fellow anthropologist A.R. Radcliffe-Brown: "human beings are connected by a complex network of social relations. I use the term 'social structure' to denote this network of actually existing relations." (1940, 2) According to Radcliffe-Brown, we are interested in "*actually existing relations.*" In contrast, Nadel focuses on relationships between actors insofar as they are "*playing roles* relative to each other." He hypothesizes, or assumes, that social networks follow systematic patterns that can be captured as roles. Not individual human beings, but "actors in roles" would be connected in networks. Such actors in roles can be subsumed in categories with systematic patterns of relationships between them.

Network scholars follow partly Radcliffe-Brown's and partly Nadel's position. Many formal analytical methods and empirical studies compare individual actors with regard to their centrality in a network, or they analyze the density or connectivity of networks. Ronald Burt (1980) characterizes this first tradition, which follows Radcliffe-Brown, as "relational." The second tradition, in contrast, adopts a "positional approach" and pursues Nadel's ambition of reconstructing systematic patterns, or role structures, in networks. It identifies categories of actors and reconstructs the systematic webs of relationships between them. While there is nothing wrong with the first approach, this chapter shows how the second approach makes for a wider perspective on network structures. It starts from the general idea that networks are patterned by categories:

Assumption 5.1. Social networks are characterized by systematic patterns between categories of actors.

Table 5.1 Some role constellations in blockmodels

(a) Core–Periphery			(b) Factions			(c) Isolation			(d) Heterophily			(e) Deference		
	P	Q		P	Q		P	Q		P	Q		P	Q
P	1	1	P	1	0	P	1	0	P	0	1	P	0	0
Q	1	0	Q	0	1	Q	0	0	Q	1	0	Q	1	0

Note: Cell entries signify observed distributions of ties from one position (row) to another one (column). Few or no ties are marked as 0; many ties or full connectivity are marked as 1, e.g., when all members of a clique are connected to each other.

We can catch rudimentary glimpses of roles in the network embeddedness of individuals, as measured in egocentric network surveys. "Brokers" are characterized by a high number of unrelated ties (to disparate social contexts). Isolates report few personal relationships. The egocentric networks of clique members show high degrees of closure. Looking at alters gives us additional information: Is the actor only connected to her family, or to high-ranking managers, politicians, and public servants? However, we can only scratch the surface of roles and of systematic network patterns with egocentric network data. These give us information about the local social environment of individuals, not about the overall network patterns of relationships.

Things look different in the central method of the positional approach: blockmodel analysis, as developed by Harrison White and his students in the 1970s (White and Breiger 1976; White, Boorman, and Breiger 1976; see sections 2.3 and 2.8.c). Here, actors are categorized on the basis of the similarity of their social relationships to others. Actors within one category may, but need not, be connected to each other. In patronage structure, for example, clients regularly exchange material goods and services with their patrons, but not with each other. The patrons, in contrast, have commercial ties to each other and to their clients. The systematic pattern of such a structure follows a core–periphery structure as displayed in table 5.1 (a), with the patrons marked as (P) and the clients marked as (Q).

Blockmodel analysis is able to decipher a number of different structural constellations, for example, highly connected factions with few ties between them (b).[1] Constellation (c) sets a cohesive subgroup apart from isolated outsiders. This resembles the established–outsiders configuration studied by

---

[1] I follow White, Boorman, and Breiger (1976, 742, 744) in the interpretation of some of the constellations sketched here.

Elias and Scotson ([1965] 1994; see sections 4.1 and 4.7). Some constellations display little or no connectivity within categories. Two categories can be tied to each other, but not among themselves (d). Heterosexual marriages are a case in point, since they always link people of different genders. Here, as elsewhere, we do not consider social relationships per se, but always specific "types of relationships" (or "types of tie"). Many friendships run between members of the same gender. But love and marriage are still, by and large, reserved for cross-gender relationships (see chapter 6).

This also holds for constellations of deference (e). Here, a subordinate category (Q) obeys and/or admires a superordinate category (P). Such deference is a typical correlate of patronage structures. Thus, (a) and (e) may map the same social constellation, with regard to different types of tie and different patterns of communication between actors: (a) represents the exchange of material goods, while (e) captures deference or loyalty. (e) is also the only constellation with directed, asymmetric ties. Many more constellations of directed types of relationships are possible. Also, I only considered simple models with two categories here. Empirically, blockmodeling constructs more complex models with typically four to eight categories.

Following Nadel, blockmodel analyses interpret the identified categories as *roles* the actors play in relation to each other (White, Boorman, and Breiger 1976, 731ff; Boorman and White 1976; Winship and Mandel 1983). Roles are here seen as tied to other roles in typical ways. The director of a school resembles other directors in her connections to students, parents, teachers, and administrative boards—even without much contact to other directors. For this, we have to distinguish relationships by their type (colleagues, superordinacy, stewardship, etc.), and by their direction. Robert Merton terms the bundle of social relationships of school directors to other roles their "role-set" (1957).

Blockmodel analysis builds on the graph theoretical concept of *structural equivalence* (Lorrain and White 1971). In the original, rather strict formulation, two actors are structurally equivalent if connected by the same ties to the same actors. Thus, the students in a class would be structurally equivalent due to their ties to the same teachers and co-students. This requirement is almost never met in practice, and impractical if strictly enforced. In the various algorithms of blockmodel analysis, a looser definition is applied: actors are structurally equivalent when resembling each other in their connections to categories of actors (rather than to singular actors).

Conjecture 5.1. Roles lead to patterns of structural equivalence in social networks. These can be reconstructed through blockmodel analysis.

Originally, the concept of structural equivalence developed out of the analysis of kinship structures. Categories like "mother" or "uncle" mark structurally equivalent positions among relatives. Subsequently, studies in quite different network contexts fruitfully applied the method of blockmodel analysis: the networks of friendship and conflict among novices in a monastery (White, Boorman, and Breiger 1976, 749ff), the exchange of women in Oceania (Bearman 1997), networks of power in community elites (Breiger 1979, 34ff; Pappi and Kappelhoff 1984; Padgett and Ansell 1993), the social structure among artists (Anheier, Gerhards, and Romo 1995), and many more. In all of these instances, the blockmodels map a fair share of the observed relationships across various types. However, blockmodels remain inductive, since they do not start from ideas about what kinds of structure they will find. This makes it hard to test conjectures.

In spite of the insights produced with blockmodels, network research leaves fundamental questions without answers: What is the reality of the observed role categories? Do they feature manifestly in the social world, recognized by the actors at play? Or do they remain latent, to be reconstructed by sociological observers? And where do they come from? In line with classical structuralism, blockmodel analysis only tackles the *structure* of social relationships. But why are networks patterned by structural equivalence? Do roles like patron or outsider emerge endogenously, in a "self-organization" of networks? Or do they follow cultural models that prescribe a division of labor, or the stratification by prestige? We can diagnose the structuration of kinship networks by institutionalized models—especially where rules for or against marriages between kinship categories hold (incest taboo, marriages between cross cousins). The studies of networks among artists, political elites, and novices in a monastery assume that the role categories and relationship patterns result in part from the interaction in the network. How can we conceptualize the relation between cultural rules and the emergence of role structures in interaction?

At this point, network research suffers from a lack of engagement with the theoretical discussions of roles and institutions in sociology. The following sections examine various approaches with regard to their fruitfulness for a relational conceptualization of roles and institutions.

5.2. Normative expectations

Roles were established as a key concept in sociology by a heterogeneous group of authors including Ralph Linton, Robert Merton, Talcott Parsons, and Ralf Dahrendorf. In spite of theoretical differences, they advance a relatively similar understanding of roles—a *normative role concept*. First, roles denoted recurrent patterns of behavior in different social situations. This includes kinship roles, as well as status or positions in formal social contexts. The concepts of "status" (Linton) and "position" (Dahrendorf) stand for the patterns of behavior in *comparison* to other status or positions. "Roles," in contrast, cover the systematic patterns of behavior *between* positions—in social relations or relationships (Linton 1936; Dahrendorf 1968, 35ff). The role concept is devised relationally: The *position* of "mother" marks it as different from the positions of "father" or "child." But the *role* of mother covers the typical social relationships to these other kinship categories.

Linton and Merton subsume both the typical behavior in a position and the expectations tied to it under the concept of roles. Dahrendorf formulates more succinctly and parsimoniously: for him, roles are expectations that the occupants of positions fulfill to different degrees and in various ways (1968, 36). Here, roles do not cover the patterns of behavior themselves, but the cause for these patterns. The role of a teacher does not consist in the similarity in behavior, but in the social norms underlying it. These norms are relational since they pertain to the behavior of actors in different positions *in relation* to each other (not of individuals seen in isolation). With this relational conceptualization, and with the focus on expectations, I can directly incorporate this approach into the theoretical perspective advanced here:

> Definition 5.1. Roles consist of expectations about the relational behavior in social positions toward other positions.

If role expectations concern relational behavior, these do not lead to behavioral patterns of isolated individuals, such as "I expect Aunt Frida to wear another spectacular hat for the wedding." Rather, the expectations concern what happens *between* actors. In the framework of this book, these expectations make for patterns of communication. These are governed by prescriptions about how incumbents of particular positions—say, a bride and her aunt—behave toward each other. This leads us back to the general interplay of communication patterns and of the meaning structure of a network, as laid out in chapter 2. Roles become part of the relational expectations in social relationships and networks that result from, and make for, patterns of communication.

> Conjecture 5.2. Role expectations make for empirically observable patterns of communication, and for a similarity of behavior in social positions.

As bundles of expectations, roles have to be "known" by the actors involved and inscribed in communication. We are not dealing with ideal-typical groupings of behavioral patterns. These could resemble each other coincidentally or out of a similarity in cultural background. Rather, behavioral patterns have to follow meaningful orientations—expectations tied to role categories. They are real types rather than ideal types (McKinney 1969). For this, roles have to feature manifestly, and to have a particular meaning, in the social context at hand. They have to be known by the actors involved (subjective meaning) and communicated accordingly (communicative meaning; see section 2.2). Following Nadel, roles carry *names* standing "directly for the role norm" (1957, 33). We can also think of other symbols, for example, forms of clothing that are specific and appropriate for a position, like a bride's wedding dress. In any case, it has to be possible to represent role expectations in communication, and to refer to them explicitly.

> Conjecture 5.3. Roles and positions are part of the manifest structure of meaning in a social context. They are represented with symbols like the name of a position.

In this sense, roles are meaningfully constructed points of orientations in communication. They structure communication between actors, and their social relationships, as normative expectations. With the strong connection to norms, Linton, Merton, and Dahrendorf link individual behavior and the communication between actors to the structures of society and to cultural patterns. According to Dahrendorf, roles are *"quasi-objective* complexes of prescriptions for behavior which are in principle independent of the individual" and "defined and redefined not by any individual, but by society" (1968, 37; original emphasis). Dahrendorf remains vague about the precise construction of roles as part of the structure of society. Some roles are inscribed in formal organization, like that of a school director. Kinship roles, in contrast, are part of cultural patterns. Roles of mothers, sons, or aunts are little formalized. They follow the cultural repertoires of nations, ethnic groups, class, or social milieus.[2]

Conjecture 5.4. Role expectations can derive from formal organization or from cultural patterns.

In both instances, role expectations are "imported" from wider social or cultural structures into the interaction between individuals. According to Talcott Parsons, role expectations develop out of the situation of "double contingency" (Parsons et al. [1951] 1959, 16; see section 2.4). In interaction, the consequences of every action depend on the reactions of others. Actors behave on the basis of expectations about the expectations of others, all of which are contingent. According to Parsons, the inherent uncertainty of interaction can only be alleviated through the "generalization" of expectations across situations, and through the "stability of meaning" of "conventions" that actors adhere to. Action systems therefore presuppose "shared symbolic systems"—culture in the sense of section 2.2—as well as normative expectations that conventions will mostly be complied with.

Parsons argues that roles draw on institutionalized culture. Expectations concerning the appropriate behavior in social positions are generalized in cultural patterns, and then applied to particular persons in interaction. "Relational institutions" are used to fix the roles of actors in a social system in relation to each other (Parsons [1951] 1964, 51f). Alter and ego are allocated to positions, this leads to relational expectations about their behavior toward the other, their actions become predictable (we can "count on" specific reactions), and the double contingency of action is reduced to the adherence to culturally defined role expectations.

The communication theoretical perspective does not locate these expectations subjectively in the heads of alter and ego (see chapter 8). Rather, such expectations are inscribed in communication (Luhmann [1984] 1995, 96f). We can trace them empirically in three ways: (1) in the communication of rules and claims, (2) in the observable regularity of behavior, and (3) in sanctions reacting to deviation.

---

[2] Of course, organizational roles themselves are subject to cultural imprint (Meyer and Rowan 1977).

5.3. Playing and making roles

In reaction and opposition to the normative approach, symbolic interactionists developed an *interpretive role concept*. Erving Goffman and Ralph Turner focus on the negotiation of role expectations in interaction. Following the interactionist tradition, roles become part of the "definition of the situation" (Goffman [1959] 1990, 20f; see section 2.2). This definition emerges in interaction, determined neither by cultural or societal prescriptions nor by the subjective dispositions of the actors involved. Goffman's formulations, in particular, are compatible with Luhmann's concept of communication: the definition of the situation does not rest on the actual agreement of subjective orientations but serves as an "interactional *modus vivendi*" (Goffman [1959] 1990, 21; original emphasis; see section 8.2). It results from previous communication and effectively structures future communication.

Against this backdrop, the interpretive role concept stresses the leeway of interaction to choose and deploy role categories. In "role-playing," roles are not merely followed, but selectively activated in communication. "Role-making" even allows for the creative shaping of roles (Turner 1962, 21f). The role of a teacher or, less formally, that of a gentleman can be performed quite differently. In turn, ego adapts her own behavior ("self-role") in reaction to the role taken by alter ("other-role"; Turner 1962, 23).

These formulations can easily be incorporated into the perspective offered here: communication can arrange, and modify, the role categories available as part of the cultural repertoire, or as set by formal organization. Following Goffman, and parallel to the formulations about relationship frames in section 2.7, we can term the cultural prescriptions for expectations by positions "role frames," as analytically distinct from the actual roles established between actors. Again, culture provides "frames" that are used to answer the question "what is going on?" between actors in a given situation. Through role-taking, as conceptualized by Turner (1962), such culturally available role frames become part of the "definition of the situation." Even prescribed roles have to be communicated, with symbols such as the name of the role, appropriate clothing, or forms of address. As with relationship frames, communication is routinely checked for signs as to whether a particular role frame is enacted.

In a sense, the careful reference to, and communication in line with, such role frames is part of "relational work" (Zelizer 2005; Bandelj 2012; see section 1.4.c). At the same time, smaller or bigger deviations from established role patterns remain possible (role-making). Role frames are adopted, confirmed, modified, or switched in the course of communication. All of this forms part of the continuous management and negotiation of expectations in communication, which determines and modifies the social relationship between alter and ego.

> Assumption 5.2. Communication determines the role expectations between the actors involved. It structures social relationships partly by building cultural and organizational prescriptions ("role frames"), and partly by deviating from them.

5.4. Plasticity and tools

In Germany, a third school of sociological theory developed a particular notion of social roles: *philosophical anthropology* around Helmuth Plessner, Arnold Gehlen, and Heinrich Popitz. Coincidentally, it advanced the concept of institutions picked up and disseminated by Peter Berger and Thomas Luckmann. Unfortunately, many of the key works of philosophical anthropology, some of them relevant for my endeavor, have not been translated into English.

Inspired by the philosopher Max Scheler, Helmuth Plessner and Arnold Gehlen became sociologists interested in how a particular perspective on human beings makes sense of social and cultural structures. Roles here constitute a response to the general problem of humanity. Humans are seen as a "defective life form," characterized by "excentric positionality" and "plasticity" (Gehlen [1940] 1988; Plessner 1969). These concepts imply that human beings are animals that lack a natural connection (based on instincts) to their environment and to each other. They have to establish artificial connections based on their intellect, and on cultural forms.

This reminds us of Parsons with his focus on the uncertainty of interaction. Indeed, philosophical anthropologists assign similar tasks to roles (and to institutions). Roles are conceptualized as "prescriptions for order," as "social anthropological support for human beings" (Claessens 1963, 13).[3] Norms

---

[3] Here and in the following, I translate quotes from the original German where no publications in English are available.

in general respond to the "plasticity" and the "need for shaping" of humans and make for their "fixation" in social structures (Popitz 2006, 63f, 90, 92). They build on the typification of actions, situations, and categories for persons (Popitz 2006, 65f; Berger and Luckmann [1966] 1991, 45ff). Roles entail expectations for these categories, and behavior is assessed against these expectations (Claessens 1963, 12; Popitz 2006, 85ff).

Consequentially, human beings are not related to each other in social structures as singular persons, but as members of categories (Popitz [1967] 1972, 14ff; 2006, 98f). In agreement with blockmodel analysis, Popitz writes of role structures as "relationships and constellations of relationships," held together by the norms of behavior attached to social categories (2006, 99). Philosophical anthropology devises roles primarily as norms for the individuals to conform to (with a certain leeway; Plessner [1960] 1985, 237f; Bahrdt 1961, 10f). But roles are not externally prescribed to human beings (or their interactions) by society. Rather, they are part of human life as in need of fixation in social structures.

Popitz also hints at social structures as networks realized in communication. He writes that norms have to applied in at least triadic social constellations (2006, 70f). Ego's response to alter violating expectations remains a "private affair." This changes with reactions by third parties—by the "group public." They can promote ego's response from individual revenge to sanctions of a general norm. Of course, this requires that third parties learn about the deviant behavior, for example, in the form of gossip (with ego talking about alter with tertius; Bergmann 1987, 61ff).

In summary, philosophical anthropology argues that norms (including roles) are activated and sanctioned in communication. The validity of norms—the reliability of expectations—is checked in constellations with at least three actors. In accordance, Gesa Lindemann emphasizes the role of third actors for the application of cultural patterns (2009, 226ff). Of course, we can always formulate normative claims in dyadic relationships. But their validity requires the embedding of dyads in larger network contexts, and at least the potential participation of third actors (with their reactions both expected and predictable). Hence, roles should not be thought of as ordering dyadic relationships, but as ordering social networks (and relationships in them).

Conjecture 5.5. The validity and the application of role expectations, like those of other norms, rely on the embeddedness of social relationships in networks.

Surprisingly, philosophical anthropologists rarely refer to institutions in their writing about roles. Arnold Gehlen offers the first systematic conceptualization of institutions in his book *Urmensch und Spätkultur* (1956; no English translation yet; the title roughly translates as: "original humans and contemporary culture"). According to Gehlen, all institutions develop from a "fundamental need for principles and stability" (1956, 49, 109). Like tools, they amplify the defective life form of human beings, underdetermined by their instincts, and in need of orientation and constraint (Gehlen 1956, 11ff). Institutions step into the "gap [. . .] of a lack of cohesion between human beings" (1956, 178). With regard to their function, institutions seem to closely resemble roles.

Institutions have to be relatively stable to order the social world symbolically (Gehlen 1956, 22, 100). Also, they should be clearly arranged and lend themselves to easy transportation from one social context to others (Gehlen 1956, 44, 94ff). Finally, they are tied to guiding ideas ("idées directrices"; Gehlen 1956, 100, 201, 287f; Rehberg 1994, 65ff). These suggest an orientation to the institutions, while remaining relatively vague—not unlike archaic myths. For example, the modern institution of marriage comes with the guiding ideas of romantic love and of communion for life.

Some institutions, like marriage, offer "recipe knowledge of the workings of human relationships" (Berger and Luckmann [1966] 1991, 57). Communication can build on this knowledge and on the expectations tied to it. This stabilizes social relationships—by fixating actors in these relationships and significantly reducing the double contingency of social interaction. As I will elaborate further, these arguments cover the relationship frames from section 2.7—they are institutions in the sense of philosophical anthropology. Of course, the normative expectations tied to these institutions will only partly be met, and communication can decide which institutions to implement, just as human beings can pick from their toolboxes (Swidler 1986). The interpretive role concept has stressed this wiggle room and the creativity of actors to tweak available role frames. However, institutions only reduce the inherent uncertainty of communication—deriving from the excentric positionality and the plasticity of human beings—if communication by and large follows their prescriptions.

Definition 5.2. Relational institutions provide models for the behavior in social relationships. These are tied to guiding ideas and stabilize constellations of relationships.

Language plays a special role here. Rules and guiding ideas build on linguistic codes. With its generalization and applicability across social situations, we can see language itself as an institution (Gehlen 1956, 195; Berger and Luckmann [1966] 1991, 85ff). Or rather, it forms a meta-institution, with other institutions based on linguistic organization (Schelsky 1970, 26).

> Assumption 5.3. Institutions are organized linguistically, since they consist of generalized expectations and of guiding ideas that have to be communicated.

5.5. Institutionalization

Roles and institutions were only brought together conceptually by Peter Berger and Thomas Luckmann. Their book *The Social Construction of Reality* ([1966] 1991) combines philosophical anthropology with Alfred Schütz's sociology of knowledge, popularizing ideas from these approaches internationally. Their focus lies on the process of institutionalization, the sedimentation of forms of meaning into cultural knowledge that actors draw on.

According to Berger and Luckmann, institutions are cultural patterns for the production of actions, for the interpretation of situations, and for the behavior between categories of persons ([1966] 1991, 90ff). Roles consist of expectations tied to categories of persons. They are based on institutions that provide rules for the behavior between categories. For example, the institution of heterosexual marriage comes with expectations about the behavior between wife and husband. Institutions are blueprints for relationships, whereas roles represent types of actors in relation to each other. However, this does not hold for all kinds of institutions, as I will argue further.

Where do institutionalized expectations come from? Following Parsons, reciprocal expectations about actions and dispositions develop as soon as an "organized system of interaction" stabilizes between alter and ego (Parsons et al. [1951] 1959, 19; see also Parsons 1961, 41f). More precisely, these expectations stabilize and organize the system of interaction. Role expectations, now, consist of a generalization of such expectations of behavior (a) across different people ("all of those in the status of ego"), and (b) across different situations of specific interaction. For Parsons, the notion of "institutionalization" denotes the implementation of relatively stable cultural patterns in social systems, with these gaining structure and stability in the process (Parsons 1961, 36).

Berger and Luckmann, and also Niklas Luhmann (1970), call the converse processes of generalization and crystallization of cultural patterns from concrete social interaction "institutionalization." Luhmann argues that behavioral expectations develop and crystallize inevitably in communication: every communicative event offers and/or provisionally stabilizes meaningful content, leading to expectations about subsequent communication. It is impossible to question, and to exhaustively negotiate, all the meaning communicated, particularly if more than two actors are involved (Luhmann 1970, 31f). With no objections voiced, a consensus about the expectations at play is assumed. Meaning gets validated in this way. A "repertoire of assumed matters of course without actual communication" develops. Luhmann here connects institutions with communication as a supra-personal process (before his turn to the concept of communication in the 1980s). Institutions arise out of the need to reduce social complexity, as an unintended consequence in the process of communication. They rely less on the "actually meant" or "actually intended" of the actors involved than on the "supposedly meant"—on the presumption that actors share the crystallized matters of course and base their subsequent actions on them.

Luhmann even shortly hints at the importance of third actors for the process of institutionalization. First, more than two actors in an encounter make it hard to negotiate definitions of the situation in detail (Luhmann 1970, 31). Secondly, the stabilized patterns of expectations are generalized by extending the presumption of consensus to absent parties and to anonymous publics (1970, 31f, 40). However, Luhmann addresses how communication develops expectations in general, without a specific concern for cultural models. Here, the notion of institutions is reserved for cultural patterns that are available in multiple contexts. Luhmann does not use this more specific concept of institutions. Hence, he does not deal with the crucial step from stabilized expectations in a given social constellation to the general availability of cultural patterns across contexts.

For institutionalization proper, we first need *generalization*, as Parsons argues (see earlier discussion): expectations have to be extended to different situations and persons. These have to be seen as comparable. Paying for items in a particular shop is regarded as similar to that in other shops, just as the salesperson is equated in her role with other salespersons. Hence, the generalization of expectations to institutions rests on typifying situations and persons as similar (Berger and Luckmann [1966] 1991, 65ff).

Berger and Luckmann offer an example that sounds brutal today ([1966] 1991, 90): An uncle thrashing his nephew does not necessarily relate to institutionalized expectations. The thrashing is still not an institution if performed regularly and predictably by the uncle. For institutionalization, this performance has to be typified and systematically related to the two categories of uncle and nephew. Then, we no longer have a singular human hitting another human being (who happens to be his relative). Instead, he performs the typical action of thrashing *as uncle*, on his nephew (not only as nephew, but *as his nephew*). He engages in "nephew-thrashing."

In this vein, behavior has to be observed as cognitively typical and as normatively appropriate across a number of social contexts, and it has to be termed (typified) accordingly. The thrashing of nephews thus becomes an institution that can be transferred to different social situations ("generalized" and "transported"; see earlier discussion). This goes beyond the mere forming of habits:

> Institutionalization occurs whenever there is a reciprocal typification of habitualized actions by types of actors. [. . .] the institution itself typifies individual actors as well as individual actions. The institution posits that actions of type X will be performed by actors of type X. (Berger and Luckmann [1966] 1991, 72)

Again, third parties play an important role: between two actors (alter and ego), an abundance of expectations arise. These characterize their relationship. Only by bringing third actors in, and by communicating typified expectations to third actors (e.g., children), these expectations become institutions. For these third actors, in particular for children, these expectations are not negotiated, but encountered as given, as nonnegotiable piecemeal patterns. They become "objective meaning" (Schütz [1932] 1967) and are experienced as "objective reality" (Berger and Luckmann [1966] 1991, 77ff).

> Conjecture 5.6. Institutions develop out of the typification and generalization of behaviors. This gives them validity in a social network (beyond their original dyadic context).

5.6. Actors and relationship patterns

Building on Berger and Luckmann, sociological neo-institutionalism conceptualizes institutions as stable and binding rules (Meyer and Rowan 1977, 341). These rules affect actors "as facts which must be taken into account" and that constrain their autonomy. In this vein, formal organizations and the relationships between them are imprinted by institutions. According to Paul DiMaggio and Walter Powell, institutions spring from the inherent uncertainty of actors in a field (1983). These orient to each other and adopt the strategies and forms of organizations they observe around them. From the wide variety of neo-institutionalist arguments, I consider only two aspects here: (1) the institution of modern actorhood, as elaborated by John Meyer and Ronald Jepperson, and (2) DiMaggio's combination of neo-institutionalism with network research, in particular blockmodeling. Both connect to the themes of this book, whereas the neo-institutionalist arguments about formal organization and organizational fields lie outside of its focus.

Regarding the first aspect, up to now, I have only examined the effects of institutions and roles on the social relationships and networks *between* actors. Meyer and Jepperson argue that the constitution of individual and corporate actors themselves—their capacity to enter social relationships as an entity with expectations and expectations of expectations attached to it— rests on cultural rules (2000, 105). Accordingly, the status of actors is culturally constructed in modernity. We attribute clear boundaries and purposes to actors, as well as the capacity to act consistently on the basis of internal dispositions.

The most important agents of modern society, individuals, formal organizations, and nation-states, resemble each other in this regard (Meyer and Jepperson 2000, 111f). All of them are seen as capable of acting rationally, and as responsible for the consequences of their behavior. For formal organizations and nation-states, additional rules suggest particular internal structures: clear hierarchies of command, functional differentiation or division of powers, etc. Actors are recognized in their field to the extent that their structure and observable behavior follow these prescribed rules of actorhood. In the perspective adopted here, the communication in a network decides on the basis of cultural models: Which entities feature as actors in the network?

Conjecture 5.7. The status of actors is allocated in the communication in a network on the basis of cultural models (institutions) of actorhood.

Regarding the second aspect, in contrast to Meyer's more culturalist work, DiMaggio connects institutions with network research. He came into contact with Harrison White and blockmodeling as a doctoral student at Harvard University in the 1970s. DiMaggio argues for the combination of cultural and relational analysis through the concepts of role and institution (1991, 1992). The relational pattering of networks follows the cultural models of roles. Roles are thus structural and cultural: structurally they correspond to network patterns, while culturally they are more or less institutionalized in role frames (see section 5.3) and part of cultural repertoires (DiMaggio 1992, 119ff). Cultural models or institutions emerge "iso-morphically" as ordering patterns in the interaction in fields (DiMaggio and Powell 1983).

According to DiMaggio, we can reconstruct institutions by way of blockmodeling the networks of actors in a field (1986). For example, if Broadway theaters assume leading positions in a network with other stages, this points to the institutional ordering of the field. DiMaggio's intertwined attention to relational and cultural aspects falls in line with relational sociology, as I discuss in the next section. Let me only register here that the role relations in blockmodel analysis can be interpreted as expressing cultural patterns.

Conjecture 5.8. Institutionalized role frames make for patterns of structural equivalence between actors (bearers of roles).

5.7. Relational sociology

Relational sociology in general traces social phenomena to underlying network patterns, and it models these networks as intricately interwoven with meaning (Pachucki and Breiger 2010; Mische 2011; Fuhse 2015b; McLean 2017). As I argue in chapter 2, networks of social relationships consist of relational expectations concerning the behavior of particular actors toward each other. These, like all expectations, develop in the course of communication and structure it in turn. Consequently, we have to relate roles and institutions to the interplay of social networks and meaning (in the process of communication).

DiMaggio offers important building blocks for the incorporation of roles and institutions into relational sociology. He sees the pattering of social networks in structurally equivalent positions as following institutionalized rules (see section 5.6). White conceptualizes institutions as stabilized socio-cultural formations developing out of the control attempts of identities (1992, 116). This stabilization builds on crystallized structures of meaning like stories or values in a particular context (1992, 136ff). However, White does not distinguish between network constellations and the cultural forms stabilizing them. For him, social formations, like the caste structure in an Indian village, or the prestige hierarchy of departments at a U.S. university, are themselves institutions. In line with most theoretical approaches discussed, I only designate the cultural patterns underlying these formations—the "recipe knowledge" (Berger and Luckmann) or "symbolic blueprints" (Portes 2010, 55)—as institutions.

In relational sociology, this line of reasoning is most prominently and clearly advocated by John Mohr, a former doctoral student of DiMaggio. Mohr studied the treatment of various categories of needy people ("Soldiers, Mothers, and Tramps," among others) by charity organizations in New York in the early 20th century (1994; Mohr and Duquenne 1997). These analyses aim at reconstructing the "institutional logics" in the welfare sector, assuming "non-determinative, mutually constitutive relationships [. . .] between symbolic orders and material practices" (Mohr and Duquenne 1997, 306, 308). Social practices (or communication) constitute symbolic orders and are in turn shaped by them. Mohr and Vincent Duquenne term this interplay of cultural patterns and social structures a "duality" (1997, 308ff; Mohr 2000, 62ff; Mohr and White 2008, 490f; Breiger 1974, 2010).

Mohr now reconstructs specifically *relational* aspects of this symbolic order, by asking for the (similar or different) treatment of social categories of people. He coins these symbolic orders "meaning structures," in a slightly different vein than I do in chapter 2: Mohr's meaning structure forms part of cultural repertoires, whereas I use the term to denote the patterns of meaning inscribed in social networks. Cultural patterns (Mohr's "meaning structures") are increasingly examined with the formal methods of network analysis (Carley 1994; Mohr 1998; Yeung 2005; Light 2014; Rule, Cointet, and Bearman 2015; Fuhse et al. 2020). The various studies in this area start from Ferdinand de Saussure's assertion that symbolic systems (like language) consist of relations between signs, words, categories, and other forms of meaning (de Saussure [1916] 2013).

As philosophical anthropologists argue, language plays an important role in cultural patterns (assumption 5.3). For example, Mohr's field of charities is organized by linguistic categories underlying the practices of treating individual needy people similarly or differently (Mohr 1994, 333f). White sees processes in networks as organized linguistically (1992, 133ff; 1995b, 706ff; Fontdevila 2010). This includes the attribution of social processes to actors in everyday "standard story-telling" (Tilly 2002, 8f, 26f; see sections 2.4 and 8.5). Following Meyer and Jepperson, this construction of actors builds on culture-specific and historically changing models of actorhood (see section 5.6).

5.8. Synthesis

In this and in the following section, I pull the various threads from the previous sections together. This section sketches the basic interplay of institutions and roles in networks. The next section applies this perspective to various forms of meaning and network patterns discussed in the previous chapters: groups, collective actors, social categories, and relationship frames.

Institutions are cultural models that provide piecemeal sets of expectations that could apply in a wide array of situations. In line with the framework from chapter 2, they are part of the wider culture that provides institutionalized forms of meaning for communication in relationships and networks. Counter to White, I accord this wider cultural context a relative autonomy from networks (Emirbayer and Goodwin 1994). Culture develops in networks, but also through other channels, including mass media, education systems, markets, and state bureaucracies.

We can term all linguistic forms and other symbols (e.g., traffic signs) "institutions" that help bring structure and predictability into communication. For relational sociology, specifically *relational institutions* are of prime interest. In contrast to Parsons, I take this concept to include only *cultural models for the relations between actors* (not for objects or concepts). This pertains to categories of (1) actors, (2) roles, (3) social relationships, and (4) patterns of relationships (networks). These various aspects of social structures are not independent: social relationships and networks as relational expectations rely on the attribution of action and relational dispositions to entities. For this, entities have to be constructed as actors that are related to others in social relationships.

Definition 5.2. Relational institutions are cultural models for organizing social relationships and their network patterns. These include role frames, relationship frames, social categories, and models of actorhood.

Hence, I do not offer an encompassing concept of institutions from the perspective of network research. Rather, one part of what is conceptualized as institutions in philosophical anthropology and in neo-institutionalism is here subsumed under the more specific term "relational institution." This allows for the empirical study of institutions by way of network research, insofar as they concern social relationships and networks. But not all institutions are "relational" and organize social relationships. Think of traffic rules or of scientific theories. As institutions, these provide blueprints for behavior in disparate situations (stop before crossing red light traffic signals; disaggregate social phenomena into individual, utility-maximizing actions). But these blueprints hold for everyone involved indiscriminately. While they structure interaction, they do not directly influence patterns of social relationships.

Surprisingly, few sociological approaches link roles and institutions systematically. The few exceptions include Berger and Luckmann, DiMaggio, and Mohr. Berger and Luckmann most succinctly conceptualize institutional patterns (like that of kinship systems) as constellations of role categories ([1966] 1991, 89ff). For them, roles always form part of a "general stock of knowledge." But this underestimates the potential to deploy and shape roles creatively in interaction (see section 5.3).

The examples from network research show that networks can be imprinted by culturally available role categories, or they can develop their idiosyncratic role patterns. Unlike Berger and Luckmann, I do not term the cultural patterns for social relationships between types of persons "roles." These constitute "role frames" and fall under "relational institutions." The role concept, in contrast, stands for the actually established *systematic patterns of relational expectations in a particular network context*. Even more than the concept of relational institution, this understanding of roles links to network research. It also conforms to the normative and interpretive approaches to roles, and to philosophical anthropology. In line with Dahrendorf (and in contrast to Linton), the role concept refers to expectations rather than to patterns of behavior or communication. Following chapter 2, communication patterns follow relational expectations, just as these map and structure sequences of communication. But roles form part of the "meaning structure" of a network.

They are forms of meaning, rather than merely ideal-typical reconstructions of observable behavior. Hence, they can be infused with institutionalized culture (as depicted in figure 2.1).

However, this concept of roles leaves open whether systematic patterns of expectations derive from role frames, or whether they develop endogenously in a network. The networks of friendship and conflict among novices in a monastery (White, Boorman, and Breiger 1976, 749ff), the networks of influence in local political elites (Breiger 1979; Padgett and Ansell 1993), and the structures of prestige and interaction among artists (Anheier, Gerhards, and Romo 1995) mentioned in section 5.1 do not merely follow prescribed relationship patterns. Rather, they are emerging orders that effectively structure communication in their respective social contexts. Outside observers can detect these endogenous orders as patterns of structural equivalence.

This simplification of complex webs of relationships to structural equivalence provides a degree of orientation and predictability in the network. Hence, it tends to stabilize itself. Rivalries, the formation of groups, and the orientation to leaders give clues about what communication to expect. They start from tentative relational definitions of the situation that crystallize into solid structures, making for regularity and predictability of communication patterns. Such role structures result from exploration in interaction (Leifer 1988). They stand for the local typification of actors in the network at hand (or "field" in the sense of DiMaggio). In this sense, we could have a typical bundle of expectations for the behavior of husbands in a local community. This would be the recognizable role of "husband" in the context. If generalized, these local types become role frames that can be transferred to other contexts. The role of husband thus becomes a relational institution that structures a number of different network contexts. This requires that local role structures resemble each other to a certain extent. Structurally similar positions could then be recognized, and generalized as types across network contexts. Following Berger and Luckmann, I call this process *institutionalization*. The opposite process of applying institutionalized patterns in local contexts is *implementation* (see section 5.5).

White interprets crystallizing role structures as the result of control attempts by identities (see section 5.7). In a later article, White picks up on Luhmann's concepts of communication and of "double contingency" (White et al. 2007). But he criticizes the focus on dyadic constellations (also prevalent in the work of his student Eric Leifer, 1988). Instead, social structures respond to "multiple contingency"—the inherent uncertainty of communication in communication with more than two actors involved (White et al. 2007, 546). Following Merton, roles form part of a "role-set." For example, the position of a teacher is tied to a number of other positions (students, colleagues, teachers) in institutionalized role relations. Similarly, husbands are not only expected to behave in particular ways toward their wives, but also to their children, in-laws, neighbors, and friends. Roles and institutions do not only structure singular social relationships symbolically. They relate multiple actors to each other, thereby offering an extended reduction of uncertainty in communication. Following Popitz, Berger and Luckmann, Lindemann, and Luhmann, the social context for the emergence and the implementation of relational institutions features at least three actors: alter, ego, and third parties reacting and sanctioning behavior, validating rules, or consenting, if only by not objecting.

In the perspective adopted here, not only social relationships and networks are ordered by roles and institutions. Relationships and networks themselves are emergent structures in the *process of communication*—a "memory" of communication (Schmitt 2009). With the "tools" (Gehlen) of roles and institutions, communication simplifies itself. Hence, communication tends to develop and to conform to role patterns of structural equivalence in a field (for example, in a monastery). For this, it partly builds on institutionalized models. The relational expectations themselves are structured in dyads: communication is oriented toward the attributed dispositions for action of singular actors in relation to singular alters. But role patterns form, and relational institutions are implemented, in the context of multiple actors.

Roles and relational institutions also connect the communication in networks to larger social structures. This includes categorical inequalities (e.g., by gender or ethnicity), the hierarchies and division of labor in formal organizations, and the various roles in fields of society (Bottero and Crossley 2011). In this vein, it should become possible to study fields with regard to the role patterns at play, as DiMaggio envisions (1986; Anheier, Gerhards, and Romo 1995). However, these issues lie outside the scope of this book.

5.9. Roles and relational institutions

The previous sections remain abstract and dry, focusing on complex theoretical arguments rather than empirical examples. I now dwell on the network phenomena captured by the twin concepts of roles and institutions. These concepts offer significant extensions to the perspective laid out in chapter 2 and cover many of the phenomena discussed up to now.

(a) Groups

In a network context, such as a company or a school class, interaction can lead to the division into cohesive subgroups, each with their distinct world-view (group culture), and defining themselves against each other (group boundary / group identity; chapter 3). The members of any of these groups would occupy similar network positions, with many positive personal relationships among them. Between groups, rivalries and conflict would prevail. The positive personal relationships in groups thus correspond to the "factions" pattern (b), while conflict ties conform to the "heterophily" constellation (d) in table 5.1. Hence, cohesive subgroups are a special instance of structural equivalence, and the membership in these groups constitutes a particular role. This role character rests on the typification of group members as similar in certain regards (group loyalty, group style).

However, the group membership and the correlating relationship patterns (the roles) arise out of internal interaction. They do not have to follow culturally available models (institutions) for dividing networks into subgroups. Groups make for a sociocultural pattern of limited institutionalization. They install a local symbolic distinction between inside and outside. This distinction gains force in the interaction within the group and to the outside world. But it stays by and large on the meso-level, rather than form part of a wider cultural repertoire.

(b) Collective actorhood

Some kinds of informal groups and collective identities are institutionalized. For example, movements of political protest started in the late 18th century, with the nationalization and the parliamentarization of political conflicts (Tilly and Wood 2009). Nowadays, people easily recognize political protest as a legitimate form of contention, and they know how to organize it. Social movements are legitimate and recognizable entities in the social world, and the cultural model is frequently picked up to organize interaction and to voice political claims. Social movements have been institutionalized, like formal organizations and nation-states (Meyer and Rowan 1977; Meyer et al. 1997).

Similarly, the American street gang is an institutionalized model for staging conflicts between deprived inner-city youth, mostly from ethnic minorities. It has proliferated widely, and not only in the United States, over the last 40 years (Klein 1995). The mass media are partly to blame, transporting images of "honorable" gang behavior with cultural insignia into every corner of the world. Terrorist groups are another instance of group formation building on cultural models, diffused across the globe and copied numerous times (Hoffmann 2006). Like the institutions of social movements and street gangs, the model of a terrorist group (and of terrorist activities) was adopted widely, but also adapted to local circumstances and to specific political goals.

These institutions of collective actorhood prescribe a certain structure of roles among the members of the collectives. More importantly, they come with particular roles to play for the collective actors themselves:

- Street gangs are pitted against police authorities, families, schools, and other bearers of conventional control in their inner-city neighborhoods. More importantly, gangs fight each other for prestige, for territory, and sometimes for stakes in illicit business (e.g., the drug trade).
- Social movements engage in political claim making, but unlike parties, they remain outside of the struggle for institutionalized political power. Occupying a peripheral position in the political field, they contend with governments and administrations, while seeking allies in the political opposition and among themselves (Habermas [1992] 1996, 359ff).
- Terrorist groups can be seen as a special type of social movement (White 1989; Tilly 2003; Goodwin 2006). They try to influence politics from the outside. But they draw on violence in their movement repertoire, as opposed to more established social movements with their nonviolent voicing of claims. This brings terrorist groups into conflict with law enforcement agencies.

In a sense, then, these models for collective identities constitute social categories of actors, which brings these actors into systematic relationship patterns with each other.

A number of authors calls social movements, street gangs, and terrorist groups "network" phenomena, because of their informal and seemingly decentralized organization (though many of them are quite centralized; Arquilla and Ronfeldt 2001; Sageman 2011). Sometimes, they are even heralded as the harbingers of "network society" (Castells [1997] 2010), or of "relational society" (Donati 2015, 101ff). I prefer classifying them as *particular kinds of network constellations*, stabilized by the emergence of a group boundary, but also by adherence to the respective cultural model. And while these particular models have developed and diffused rather recently, we have no way of determining whether our current (or "coming") society features more collective identities with informal and (seemingly) decentralized organization than before (Maffesoli [1988] 1996). Proclamations of a network society are premature and lack empirical evidence.

(c) Social categories

Social categories, like ethnicity or religion, are themselves relational institutions. They form part of a wider cultural repertoire, and they come with expectations about how the members of these categories interact with each other (within and between categories; chapter 4). They prescribe roles for those categorized in relation to each other. Of course, communication always has some leeway in allocating these roles, and in shaping them. But social categories are hard to ignore, if only because they give us clues as to what to expect from each other (thus reducing social uncertainty). As institutions, social categories rest on typification and generalization, but they also vary in salience and in the expectations attached to them across contexts and over time.

Many social categories lead to structural constellations of cohesive subgroups, with high degrees of homophily (in positive personal relationships) and low heterophily. This is the common instance of a "factions" constellation (see table 5.1.b). As a case of structural equivalence, it is rather boring. Other social constellations—for example, feudal hierarchies and family relations—feature social categories with more equivalence and less cohesion. Feudal nobility (of a particular rank) is a case in point, as are the 19th-century allotment farmers in France (Marx [1852] 2005, 83ff; see section 2.8). Gender is another tricky instance of a social category of structural equivalence (Seeley 2014; see chapter 6).

The social cognition approach of Eviatar Zerubavel (1991; Brubaker 2004, 71ff) conceptualizes categories as mental schemas for observing and classifying people, often on the basis of external traits (see section 3.6). I agree that social categories make for a reduction of social complexity (and uncertainty), and that the social world can hardly do without them. But it seems more fruitful to see categories as institutions rather than as cognitive devices, because they carry very different meanings across contexts and time, and because they vary widely in their salience. This squares with Zerubavel's assertion that categories are culture dependent (1991, 61ff), and with Rogers Brubaker's argument that ethnic categories are frequently promoted by po- litical elites (2004, 150ff). Devising social categories as necessary mental simplifications of the world goes halfway to viewing them as universal and independent of social and cultural structures. But categories differ in their salience and frequently change over time.

For example, it was commonplace in Germany until the second half of the 20th century to categorize people by their adherence to Lutheran Protestantism or to Catholicism. Nowadays, this makes little difference in interaction. Why this change, if we need such categories for ordering the world cognitively? Categories should be conceptualized as performed in communication rather than subjectively thought, be it in the "doing of dif- ference" (West and Fenstermaker 1995) or in "boundary work" (Lamont and Molnár 2002). Social categories have to be enacted and reproduced in communication, like all social and cultural structures. And they come with culture-specific expectations about how to behave with and between categories.

(d) Relationship frames

Finally, relationship frames like love, friendship, and patronage are cultural models that we use to categorize our personal relationships, and to reduce the uncertainty in the communication with others. As I elaborate in section 2.7, such relationship frames are culturally variable across contexts and time. For example, our understandings of love and friendship have changed consider- ably over the last 300 years (Luhmann [1982] 1986; Giddens 1992; Jamieson 1998; Illouz 2012; see section 6.2.a). The word "love" remains, but we attach very different expectations to it. However, these are idealized frames, and any particular relationship will differ with regard to the actual expectations at play. They derive partly from the model, but also from the ongoing negoti- ations between lovers. We routinely engage in "relational work" to mark our relationships by reference to such cultural models, and to determine what exactly we can expect of each other (Bandelj 2012).

If communication between alter and ego agrees to define their relationship with a particular frame—"love," "friendship," "patronage"—this entails assuming particular roles: lovers, friends, patron/client, etc. Relationship frames are coupled to role frames. They come with particular bundles of expectations. These concern the behavior between alter and ego, but also that in relation to third parties, as I elaborate in section 2.7.b. Lovers are expected to remain faithful and to abstain from becoming intimate with others. And they have to arrange themselves with each other's friends and the family, just as these have to accommodate the partner of their friend or family member. Friends are expected to treat each other's partners, friends, and family amicably, but also to show solidarity with each other in times of trouble (e.g., in case of a conflict between one friend and her partner). Clients have to be loyal to their patrons and to abstain from offering their loyalty to other patrons.

Relationship frames, as institutions, symbolically prescribe certain relational expectations (including social relationships and network constellations), as well as patterns of communication. Following DiMaggio, they lead to constellations of structural equivalence. Blockmodel analyses should give us clues about the institutions at play in a network context. But in line with the general approach of relational sociology, we should also trace relationship frames, like all symbolic forms, in communication (McLean 2007), rather than only infer them from structural patterns.

5.10. Conclusion

Actors, relationships, and relationship patterns—all aspects of social networks seem to be imprinted and to follow institutionalized models. However, this is only partly true:

- In order for *actors* to be recognized, and to have relational expectations attached to them, they have to conform to one of the institutions of actorhood: individuals, collective actors, formal organizations, or nation-states. Meeting the institutional rule is a necessary requirement here.
- *Social relationships*, in contrast, need not comply with any one relationship frame. It would be easier to analyze social networks if the ties

all fell into neatly categorized types, and if the ties in a category by and large resembled each other. However, the communication between actors inevitably leads to relational expectations (unless it sticks entirely to formal roles like vendor/client), and these are clumsily classified through culturally available relationship frames. Just think of the many different kinds of relationship termed "friendship" (Fischer 1982b). Still, we are often not sure whether somebody is a friend to us, or just a colleague or neighbor.

- *Network patterns* may be imprinted by cultural models (like patronage structures, or the homophily within social categories). But they also result from the complex intermingling of multiple social contexts (e.g., if one's daughter's teacher is also a neighbor) and from the general indeterminacy of communication. Blockmodeling partly takes this into consideration by focusing on systematic patterns rather than on details in the actual structure. However, it does assume a homology of the role structure (and therefore of the underlying relational institutions). We might have to develop methods that focus on structural equivalence but allow for the gradual intermingling of multiple institutional logics, just like exponential random graph models allow for multiple network mechanisms at once, and to varying degrees (Lusher, Koskinen, and Robins 2013).

Relational institutions bring a certain degree of order into networks of social relationships. They partly do so through roles, making for patterns of structural equivalence. But they also provide models for social relationships, they classify actors into social categories, and they qualify what kinds of actors we attach relational expectations to. However, the actual network patterns are not only influenced by institutional models, but also by opportunities for contact, by network mechanisms, and by sheer coincidence.

While relational institutions make for a symbolic ordering of social networks, partly through role patterns, they stand in tension with the process of communication and other factors driving it. We can see this by fitting institutions and roles in the general model from figure 2.1 (see figure 5.1). Relational institutions like role frames, models for actorhood, social categories, and relationship frames form part of the wider culture. Communication draws on them to establish relational expectations between actors. These include role patterns and types of tie as part of the "meaning structure" of social networks. This meaning structure tends to follow institutionalized blueprints, but only to the extent that they correspond to the actual communication patterns. As always, communication is guided by relational expectations, but also by other factors, including foci of activity like schools, workplaces, neighborhoods, and other meeting places.

Figure 5.1 deviates from the framework in figure 2.1 in two important points: First, foci of activity and relational institutions do not only imprint patterns of communication and relational expectations. They also partly result from them. Foci of activity crystallize in patterns of communication around sociospatial configurations and from the stochastic accumulation of particular contacts in specific places. More importantly for our discussion, relational institutions develop from relational expectations in the process of institutionalization. Local communication patterns thus leave their imprint on cultural repertoires.

Second, foci of activity are themselves subject to cultural models, be they schools, companies, voluntary associations, bars, cafes, etc. Almost all foci of activity adhere to such institutionalized blueprints; otherwise, people would not know what to expect or how to connect to others there (and not go there). The heterosexual singles dance venue from section 2.1 is a case in point (Beattie et al. 2005): it works as a focus of activity precisely because people expect to meet certain kinds of others there and to engage in particular activities with them—maybe even to form particular kinds of relationships.

The twin notions of institutions and roles thus help make sense of structural equivalence in networks. But they also integrate the concepts discussed in the previous chapters into a systematic framework. Cohesive groups can be seen as a special instance of role patterns of structural equivalence. Social categories and relationship frames are relational institutions, forming part

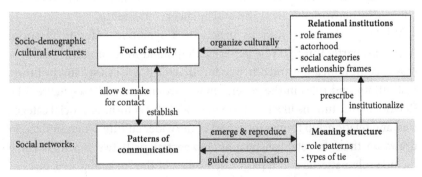

Figure 5.1 The implementation of relational institutions in social networks.

of wider cultural repertoires (subject to the dual processes of institutional-ization and implementation). However, not all institutions are relational, and not all role patterns conform to cultural models. Social reality tends to simplify itself, to make itself more manageable and to reduce uncertainty. As researchers, we have to deal not only with this simplified version of the so-cial world, or with actors' knowledge of it. Rather, we have to come to terms with the interplay between this mapping and simplification in meaning (institutions) and the complex patterns mapped, and resulting from it.

6

Love and Gender

Love and gender feature centrally in the social world, but they are rarely engaged with by general sociological theory. Talcott Parsons wrote about the family ([1943] 1954). Niklas Luhmann ([1982] 1986), Ulrich Beck and his wife Elisabeth Beck-Gernsheim (herself a renowned sociologist of the family; 1990), and Anthony Giddens (1992) published on love in modern society. Erving Goffman (1977), Pierre Bourdieu ([1998] 2002), and Randall Collins (1971, 1991) offer their views on gender. But overall, these treatises remain exceptions and peripheral in theoretical discourse. If we want to learn about love and gender, we have to turn to specialized authors like Simone de Beauvoir, Judith Butler, Raewyn Connell, and Eva Illouz. Among prominent (mostly male) general theorists, only Peter Blau (1964, 76ff), Harold Garfinkel ([1967] 1984, 116ff), and Michel Foucault ([1976] 1978) address these topics as part of their core oeuvre. Are love and gender special, in the sense of not lending themselves to systematic theory?

With its dual focus on forms of meaning (like the gender category) and social relationships (like love), relational sociology promises to deal with the matter more directly and systematically than, say, systems theory, field theory, or the theory of action. But the core authors of relational sociology—Harrison White, Charles Tilly, Mustafa Emirbayer, Ann Mische, and Nick Crossley—remain mostly silent on love and gender. In this chapter, I build on the general framework from chapters 2 and 5 to fill this gap and develop a relational sociological account of love and gender in networks. Drawing on the conceptual tools from the preceding chapters, I advance the following arguments:

(1) The social category of gender and the relationship frame of love are cultural models for organizing interaction and social relationships—they are "*relational institutions*."

(2) Both of them come with culture-specific *expectations* about how people should relate to others of the same and of the opposite gender, including the formation of love relationships.

Social Networks of Meaning and Communication. Jan Fuhse, Oxford University Press. © Oxford University Press 2022.
DOI: 10.1093/oso/9780190275433.003.0006

(3) The gender category assigns specific positions to women and men, respectively, in social life, including the public sphere, the workplace, and the family. In network terms, gender makes for patterns of *structural equivalence*, rather than for cohesive networks within categories (like ethnicity). While this clear-cut division is eroding in Western societies, it still symbolically allocates women to subordinate and domestic positions.

(4) The construction of the genders is tied to the ways in which they relate to each other—most importantly, in the relationship frames of *heterosexual love* and of *family* relations. Categories and identities correlate with network positions, and with institutionalized forms of relating. As long as we adhere to traditional, heterosexual, and asymmetric understandings of love and family, women and men are held in their symbolically and materially unequal places. But this is already changing, as institutionalized social categories and relationship frames constantly develop in their meaning and salience.

Of course, I am not able to develop a full theoretical account of love and gender in one book chapter, let alone engage adequately with the abundant and varied scholarship on the matter. I first offer (highly selective) summaries of the scholarship on gender (section 6.1) and love (section 6.2). Then I consider various attempts to link gender and networks theoretically (section 6.3) and the sexual fields approach (section 6.4). The following sections advance my relational sociological account in three steps: Love and gender are discussed as institutionalized blueprints of personal relationships (section 6.5) that make for patterns of structural equivalence (section 6.6). Then I offer three conjectures about gender-typical embeddedness of heterosexual love ties in networks of personal relationships (section 6.7). Finally, a short empirical analysis compares women and men with regard to the confidants named in the *US General Social Survey* (GSS) (section 6.8).

6.1. Gender

Modern scholarship on gender starts from this assumption: biological differences between men and women do not explain or legitimize their differential treatment in social life (Kessler and McKenna 1978, 42ff; Epstein 1988, 46ff). This has led to the switch from (biological) sex to (socially constructed) gender as the distinction between male and female. Gender is regarded as a culturally variable social category, rather than a fixed trait of human beings. However, this switch leaves a number of questions open, and the nature and the peculiarities of the gender category still a conundrum.

De Beauvoir finds "the subject matter irritating" ([1949] 1976a, 11). In contrast to the distinctions of class and of ethnicity, women are intimately connected to their "other" (men) in couples and families ([1949] 1976a, 19f; [1949] 1976b, 221ff). Like the working class, women tend to be subject to social domination and material deprivation, and their work and contributions are not recognized, valued, and rewarded as much as those of men. But class is defined as springing from economic relations. Gender, in contrast, seems to come first, providing the basis for household relations (with women dominated and in economic disadvantage), rather than vice versa. In addition, the similarity of behavior by economic status and class consciousness are rooted in cohesive interaction networks within each class (Wright and Cho 1992; Gould 1995), as is probably true for many instances of groups and social categories (see chapters 3 and 4). The two genders, in contrast, interact quite frequently and intimately with each other.[1] Gender differences in behavior and the identification with a gender cannot come from the dense interaction within a group. Accordingly, gender seems to be a social category that fundamentally structures the interaction between people, rather than emerging from it.

Of course, de Beauvoir wrote 70 years ago and offers a rather gloomy vision of gender relations. The social world has changed in the meantime but certainly has not fully eradicated gender asymmetry. I find it useful to first isolate gender (and romantic love) in its traditional version as an ideal-typical model, to see how social relations were (and could be) patterned by it. This theoretical sketch focuses on the organization of personal relationships, touching economic and political organization only in passing. Later in the chapter, I examine to what extent we still find gender differences with regard to confidant relationships in recent network data (section 6.8).

We can start by thinking of gender as a schema for classifying people (Zerubavel 1991, 46f). When encountering others, we look for cues for

---

[1] I focus on the two genders female and male in this chapter, since these constitute the prevalent categories in which people are sorted by gender. This is purely reconstructive in intent and in no way denies the right or the desirability to identify with other genders, or to not identify by gender. Diagnosing that things are in a certain way does not preclude looking for alternatives. I see diagnosis as a prerequisite for doing so.

whether they are male or female. Facial features, hair, bodily contours, clothing, posture, and names are examined for traits that are typically male or female. Conversely, most people seem keen on conforming to these types, accentuating traits that allow others to recognize the gender (Goffman 1977). Women grow their hair, apply makeup, and wear blouses, skirts, and dresses (or T-shirts, trousers, and blazers accentuating their femininity). Men cut their hair short, grow a beard, and wear suits or other typically male clothes. As a consequence, the social world is almost entirely populated by people easily classified as male or female.

This allows for gender to become a "principle of vision [...] inscribed [...] in objective divisions" that also organizes "the perception of these objective divisions" (Bourdieu [1998] 2002, 11f). The category makes us see people in a certain way, and it divides them effectively, allocating them to their appropriate positions in society. Women are regarded as "the weak sex" and as in need of protection. Their traditional tasks are centered around child care and the household. Accordingly, they were seen as less qualified when entering the workforce and public-political life, having to fight for voting rights and for equality of pay. The gender category organizes diverse realms of social life: from love and the family to the workplace and politics (Epstein 1988).

To effectively organize social life, gender has to be connected to other features of it. In the example of the Kabyle people in Algeria, Pierre Bourdieu reconstructs a symbolic universe divided into male and female ([1998] 2002, 9ff). Women are here linked to the world of "humidity" (as master concept), including cultivated nature, night, black, winter, North, cold, closed, left, on the bottom, and dominated. Men, in contrast, are symbolically connected to the master concept of "dry," and linked to South, day, summer, hot, white, right and rights, open, dominant, and sacred. While this cultural cosmos sounds alien to us, we recognize that men have the upper hand, as expected.

Students of culture have a tendency to overstate the coherence of the systems they study, and perhaps modern society allows for more wiggle room and openness. But other authors have noted, too, that male and female form part of larger repertoires of cultural practices and beliefs. For example, Deborah Tannen identifies different speaking habits of women and men ([1990] 2007, 1996). Typically, men are more oriented toward independence and autonomous problem-solving, whereas women tend to emphasize relationships and community. This resonates with Carol Gilligan's argument that women tend to develop a distinctive morality

([1982] 2016). When confronted with moral dilemmas, they often try to solve them through communication. Male morality, in contrast, is more principled and less communicative.[2]

Gender is clearly an asymmetrical distinction. In a way, it only designates the female gender and distinguishes it from the male default option (de Beauvoir [1949] 1976a, 14ff; Irigaray [1977] 1985). People are usually thought of as male, with women as deviating cases. The gender category "man" builds on the modern idea of autonomous individuals, endowed with the capacities for rational reflection and with unalienable rights, unburdened by ties to others. Women are contrasted as caring, dependent, embedded, and irrational. Unsurprisingly, they had and have to fight for the rights of autonomy and equal treatment. However, R.W. Connell shows that traditional masculinity has come under pressure. A variety of "masculinities"—different styles of dealing with the tension between what is traditionally male and the gender reflection called for by feminism—develop in response (Connell [1995] 2005]).

Evidently, gender and the ways of dealing with it are stereotypical and partly historical. The category does not hold in the same way across all times and sociocultural contexts. Following the arguments in chapter 5, we can think of gender as a cultural model, a "*relational institution*" that helps structure interaction (Martin 2004). As a social category, gender comes with a specific prescription for how to deal within and across the category (among males, among females, from man to woman, and from woman to man; see sections 4.6 and 5.9.c). The classification of people as male or female activates particular expectations—of how we are to deal with people of the opposite, or of the same, gender. Of course, these expectations are more or less salient depending on the social context. A singles bar or a wedding are more imbued with prescriptions for the genders, and a supermarket with less. Nowadays, many social contexts feature a prescription to *disregard* gender prescriptions, or to avoid gender-typical behavior, for example, at the workplace and in academia. As an institution, the expectations tied to gender are historically changing, and they depend on social contexts. For example, people in large Swedish cities tend to treat others in less gendered ways than in rural Southern Italy.

---

[2] Even the symbolic cosmos of contemporary sociology seems to be organized (in part) by the gender category. Feminist sociology is often connected to qualitative interpretation (communicatively probing the subjective meaning of its subjects) and to moral reasoning. Typically male sociology focuses on abstract theoretical arguments and technical observations through statistics.

In a similar way, Gayle Rubin writes of a "sex/gender system" (1975) that organizes social relations, and Connell uses the term "gender order" (1987). Gender as an institution has to be conceptually distinguished from the communication adhering more or less to the expectations tied to it. While it is hard to ignore institutionalized expectations, communication can always follow them to a larger or lesser extent. It can also renegotiate the expectations between the people involved, modifying or even combining the available cultural models. Finally, as Connell's research on masculinities suggests, institutions are not unitary ([1995] 2005). Contemporary culture offers us more than one way to enact femininity and masculinity in its tool-kit (Swidler 1986).

If we think of gender as an institution, we have to look for the communicative practices it enables and channels, but also for the ones reproducing or changing the expectations tied to it. Here, the notion of gender as a cognitive schema becomes impractical. Instead, gender is "done" (West and Zimmerman 1987), and has to be done, in order to remain forceful and convincing. Gender is done in the wearing of gendered clothing, in the cutting or growing of hair, and in posture and demeanor. It is done when people choose partners that conform to gender ideals relative to them: women tend to partner with men who are taller than them, older, and better educated (Goffman 1977; Kaufmann 1993). As a consequence, women appear weaker, and men—more advanced on the career ladder—typically assume the role of main breadwinner in their families. Things would look different if we sorted ourselves into couples differently. Gender is also inscribed in verbal communication: in gender-typical styles of conversation (Tannen [1990] 2007, 1996), including different propensities to interrupt each other (Smith-Lovin and Brody 1989). Also, women and men exhibit, and decode, different flirting behaviors when dating (McFarland, Jurafsky, and Rawlings 2013).

These gender-typical behaviors are astounding. But they are hardly able to make for the production and reproduction of the gender distinction on their own. French poststructuralism offers a different solution to the gender puzzle: Michel Foucault sees the two genders as part of a "norm" of heterosexuality ([1976] 1978; cf. Butler [1990] 2007). This norm is centered on the male–female couple, branching out to the family as a "network of pleasures-powers" (Foucault [1976] 1978, 10, 46, 108). The division of people into two mutually exclusive genders, and heterosexuality as their mating principle, organizes a whole complex including personal desires and pleasures, and social relationships. According to Foucault, even our feelings are imprinted by the prevalent heterosexual discourse. Therefore, we err when locating desires and pleasures in biological nature, independent of the social and cultural world, and somehow primordial to it.[3] Gender is deeply engrained in the social structures of kinship relations. And our embeddedness in the family leads us to regard the division of the sexes and the formation of heterosexual couples as natural (Wittig 1980). Within this sociocultural arrangement, women are assigned the roles of objects of desire ("sexual beings") and of motherhood ("with an obligation to reproduce"; Wittig 1982, 67).

In the second half of the 20th century, a number of feminists suggested that women have to leave this heterosexual regime through lesbianism in order to escape their subordinate position (e.g., Wittig 1980, 110). It may also be possible to change the regime itself, either by renegotiating heterosexual relationships, or by subscribing to new models like "androgynous" or "confluent" love (Cancian 1987, 105ff; Giddens 1992, 61ff, 188ff). In any case, the construction of the genders seems to be intertwined with that of love and family.

6.2. Romantic love

Turning to the topic of love, we encounter a very different kind of scholarship in the social sciences. Much of the literature focuses on historical change. It demonstrates that the modern ideal of love is indeed modern, and it traces changes of this ideal over time (Luhmann [1982] 1986; Cancian 1987, 15ff; Beck and Beck-Gernsheim [1990] 1995, 45ff; Jamieson 1998, 15ff; Illouz 2012, 18ff). I first present a rough sketch of this history of love. Then I turn to general features of modern "romantic" love, including its linkage to gender. I concentrate on heterosexual intimate ties, since this is how romantic love was traditionally defined, and still mostly is.

(a) The history of love

Most authors agree that a fundamental shift occurred with the onset of modernity and commercial capitalism in the Western world around 1800.

---

[3] Umberto Eco's book *The Name of the Rose* ([1980] 1983) and the movie *The Truman Show* (Niccol 1998) are two of the clearest literary examples of the widespread trend to regard love as somehow outside of, and opposed to, sociocultural constructions.

Before, love and marriage were firmly embedded in larger familial networks and local communities. Intimate relations were dealt with more practically, rather than thought of in emotionalized, "romantic" ways. At the beginning of the transformation, courtship was pursued within a "social web," as Eva Illouz argues, referring to the novels of Jane Austen (Illouz 2012, 27ff). Location in social strata was crucial for finding a suitable partner, and one's friends and families helped determine the qualities of prospective spouses. With the careful management of one's relations in the wider web of friends and family, lovers were not judged on the basis of their authenticity, but with regard to their ability to play their role(s) on various social stages.

Through the 19th century, the ideal of romantic choice, against social barriers and against the interests of the family, emerged, (Marriage markets" were disembedded from the economic and political realms, and prospective partners increasingly only judged on the basis of intrinsic qualities, rather than by their social place (and their ability to play their roles). Norbert Elias traces this ideal to a "romantic" resistance of the rural gentry against the refined arrangement of social identities and relations at the court of the French absolutist kings ([1969] 2006, 230ff). At the court in Versailles, aristocrats had to engage in Goffmanian presentation of identities and in managing one's relationships through careful observation and canny conversation. People figuratively, and often literally, wore masks. Fleeting sexual relationships were part of this courtly world and dealt with discreetly and casually. Elias examines one of the first formulations of romantic love in detail, in the French pastoral novel *L'Astrée* by Honoré d'Urfé, published in six parts between 1607 and 1627. D'Urfé laments the superficial and often instrumental toying with others in courtly intimate relationships. He contrasts it with ideals of authenticity and faithfulness, connected to an idealized (romanticized) rural past. The resulting ideal of romantic love does not involve masks and fleeting relationships, but deep and authentic longing. Love is a cure against the pain inflicted by the artificiality and shallowness of early modern society.

Elias thus relates the emergence of the modern ideal of love to a specific social constellation in early modern times: the conflict between traditional rural gentry and the central court society. Other writers offer different explanations. A common argument links romantic love to the rise of commercial capitalism around the turn of the 18th century. According to Francesca Cancian, intimate relationships were firmly embedded in the wider context of familial, economic, and feudal relations before 1800 (1987,

15ff). With the increasing commercialization of society, "love" crystallized as a second sphere in opposition to the anonymous and superficial relations in the economy and in public-political life. Love became "feminized": with men in charge of public and economic relations, women assumed responsibility for intimacy and for the family, now disembedded from the public community as the core family of the bourgeoisie. The gender ideals sketched earlier correspond to the new division of labor: Men had to be goal-oriented, independent problem-solvers. Women specialized in empathy and in managing familial relations.

Unsurprisingly, Niklas Luhmann's account of love does not focus on social conflict, nor on capitalism. Rather, he locates the emergence of "love as passion" in his wider diagnosis of modernity as the increasing *differentiation of functional subsystems* (including the capitalist economy, but also politics, law, science, the mass media, and the family; [1982] 1986). With the erosion of premodern feudal stratification, social relationships have become more volatile, and less prone to societal pressures. In the economy, in law, in science, and in politics, new "impersonal relationships" are coordinated through the media of money, power, law, and scientific truth (Luhmann [1982] 1986, 12ff). This development freed *personal* relationships from economic, political, and legal exigencies. Love relationships could intensify and be stylized as counterpoints to impersonal structures of society. At the same time, love builds on the modern semantics of *individualism* (Luhmann [1982] 1986, 97ff; Luhmann 1990, 107ff). It supposedly connects two individuals with their very own qualities and selves, uninhibited by social constraints and conventions.

As in the pastoral novel *L'Astrée*, love is seen as "natural" and contrasted with the artificial social world. At the same time, love is opposed to reason as the principle of a "rationalized" society (Luhmann [1982] 1986, 94ff). Economic and political relations are organized around means–ends calculations. Romantic love, in contrast, is irrational (unlike the fleeting intimate relationships at court). It does not ask for returns of efforts. And it does not choose its subjects rationally. In a sense, the belief in love becomes religious through the 19th century (Beck and Beck-Gernsheim [1990] 1995, 168ff). It is worshiped and pursued by modern individuals in spite of abundant evidence that real-life relationships cannot live up to the ideal.

However, it should be noted that we are dealing with the practices and representations of love in commercial and cultural elites. Among peasants and in the working class, love tended to be much more practical, less romanticized, and firmly embedded in the web of familial and communal relations far into the 20th century. But these, too, gradually picked up on elite ideals of romantic love and incorporated them into their relationship formation and practices.

(b) Central features

Weaving these various threads together, the modern ideal of "romantic" love looks roughly like this:

- Love is *individualized*. It focuses on the individual in all its aspects (diffusely). The whole self is immersed in a love relationship but also enabled to develop, unfold in it (Cancian 1987, 105ff).
- Romantic love is *authentic* and opposed to the artificial modernity. Love is stylized as a counterpart to the impersonalized spheres of politics, economy, law, and science.
- Love is *dyadic* and *dependable*. It connects two people exclusively, forbidding intimate relationships with others.
- Love is tied to *sexuality*. Partners engage in sexual intimacy, and sexuality is expected not to take place outside of love relationships.
- Love, unless disappointed, is destined to be *stabilized* in *marriage*. Modern love (in its traditional version) is supposed to last forever (until death). This strict monogamy has now largely been replaced with "serial monogamy." This allows people to engage in a succession of serious and monogamous relationships, one at a time.
- Love is expected to run between one man and one woman (*heterosexuality*). Homosexuality is frequently regarded as less serious and nonbinding, also sometimes as unnatural. In any case, it is running against the norm (Foucault). Over the last 100 years, the idea of romantic love has been extended to homosexual relationships, up to the rights of same-sex marriage. In the following, I discuss only the traditional heterosexual understanding of love as the blueprint against which to measure current social patterns.
- Heterosexual love institutes a *gendered division of labor*, with the woman responsible for familial relations, for child care, and for the management of the household. Men are assigned the roles of breadwinner and of managing public and economic relations.

- Since sexuality is domesticated in love relationships, and these involve a man and a woman, love bears the natural fruit of *children*. Love and marriage form the nucleus of the modern *family*.
- Romantic love *breaks through* the *social barriers* of feudal estates, class, religion, confession, and ethnicity.

Importantly, love and gender are tightly linked to each other. Romantic love comes with expectations of how the genders behave toward each other, and this partly constitutes them. If love is "feminized," as Cancian argues, then women are more tied by love then men. Men, as responsible for the economic relations of the household, are closer to the modern ideal of the independent individual. Women, in contrast, are primarily connected to their husbands and children and to the wider family. They are not free and individual but exist only as part of their households—they are dependent.

The feminist movement, the increasing incorporation of women in public-political life and in the workforce, and the sexual revolution of the 1960s have put some features of romantic love under pressure (Wouters 1998). In particular, the gender roles in love and in the family are eroding. Cancian argues for an ideal of "androgynous love" with a dual focus on commitment and self-development (1987, 105ff). Giddens diagnoses a trend to the new ideal of "confluent love" with less gender inequality and more "democracy" in intimate relations (1992, 61ff, 188ff). Confluent love could also be homosexual, and it does not have to be monogamous. Overall, it seems that full-scale sexual liberation, real gender equality in intimate relationships, and unconventional arrangements like homosexuality and polyamory remain relatively minor social phenomena at the margins of society, mostly in liberal enclaves like New York, San Francisco, Barcelona, Paris, or Berlin.

(c) Love as institution

Now, what is love? The discussion so far leads to two conclusions: First, the ideal of love is changing out of shifts in social constellations and in tandem with larger developments in society. Love is not natural, but "naturalized" as a cultural trope. Second, love can be found both in cultural constructions in novels and treatises (*L'Astrée*, Jane Austen, the many publications surveyed by Luhmann), and in actual intimate relationships. Lynn Jamieson distinguishes between the "public stories" told of love in the mass media, the actual practices in relationships, and the "private stories" about these actual relationships (1998, 10ff, 159ff). These three levels inform each other. Austen's writings build on thorough observations of the negotiations of love and marriage in Britain around 1800. And numerous people learn from novels and popular writings about courtship and intimate relationships. Luhmann writes that "love is an emotion preformed, and indeed prescribed, in literature" ([1982] 1986, 45).

Building on the arguments of chapter 5, we can term the ideal(s) of love a *"relational institution"*—a cultural model for how to construct social relationships and network patterns (see also Swidler 2001, 175ff, 196ff). Love as represented in the published stories of the mass media provides prescriptive knowledge. This cultural blueprint changes over time, and it can differ from one sociocultural context to the next. Love looks different in the early 21st century than in the 19th. And it is differently prescribed in the United States than in France or in Japan. Love can even be connected to one set of norms in the liberal milieus of San Francisco and Berlin, and to another one in rural West Virginia or Bavaria. King-To Yeung shows that "love" varies in its meaning across urban communes in the United States in the 1970s (2005). As an institution, love ideals depend on their representation and diffusion in the mass media, but also on the mutual observation and orientation in local networks.

More precisely, love is a *relationship frame*, as coined in section 2.7 (Fuhse 2013). It offers a bundle of expectations that people can draw on to structure their relationships. By adopting the frame of love, they define their tie as conforming to the culturally available ideal, at least roughly. Saying "I love you" thus translates into "I allow you to expect of me the behaviors typically associated with the frame 'love' in our socio-cultural context. And I want to expect the same of you." That is a mouthful, and not very romantic. Better stick to the short version.

Like other relationship frames, love triggers particular *role frames* when activated. For example, if two people define their relationship as a friendship, it assigns the role of "friend" to them, with particular expectations. Unlike friendship, but like patronage and many other relationship frames, romantic love comes with asymmetric role frames: lovers are assigned the male or the female part in a relationship. In marriage, these are called "husband" and "wife." The social category of gender is deeply inscribed in romantic love. Many gender-typical behaviors are not expected of women and men in relation to indiscriminate others, but only in their intimate relationships to lovers and spouses. In particular, women are traditionally expected to be tied down by their marriages and families. They are often denied jobs or promotion partly because they are not supposed to be main breadwinners, or because they are seen as less reliable due to (potentially) having children.

(d) Love as relationship

The social sciences do not only know about love from the "public stories" of novels, movies, and advice books. Love "on the ground" in actual intimate relationships has been studied extensively by way of qualitative interviews. Many studies here reveal a tension between the "public stories" about love in the mass media on the one hand, and people's "private stories" (Jamieson) on the other hand. Working with vignettes, Eva Illouz finds her respondents drawn toward "enchanted" narratives in line with the romantic ideal of love (1998). Preferred stories include "love at first sight," "fighting for love against obstacles," and "happily ever after." Actual relationships tend to be more mundane, with an emphasis on complementarity of interests and mutual understanding. Similarly, Jean-Claude Kaufmann's female informants report dreaming of a "fairytale prince," often while living in satisfying intimate relationships with less princely partners ([1999] 2008). Ann Swidler finds that her divorced respondents (both male and female) disdain the "romantic love mythology" while at the same time clinging to it (2001, 111ff). Apparently, actual relationships rarely live up to the mass-mediated trope of romantic love. The ideal remains attractive against the evidence.

Nevertheless, the institutionalized expectations connected to the love frame structure actual intimate relationships. Even if the ideals of "love at first sight" and "happily ever after" are not met, most intimate relationships remain dyadic and are at least normatively expected to be monogamous. Many of them provide empathy, albeit differently for the genders (Bearman and Parigi 2004, 541ff). Most intimate relationships are heterosexual and feature a gendered division of labor (Kaufmann 1992). Few of them break the barriers of class, religion, and ethnicity (Rosenfeld 2008). Many love ties are not as authentic and dependable and do not offer as much understanding as expected—adding to the disillusions of love, and leading many couples to divorce (Riessman 1990; Swidler 2001). Roughly one-half of marriages are divorced in the Organization for Economic Cooperation and Development countries (OECD; with the United States slightly above, and the EU slightly below the 50 percent line). The other half of marriages holds "till death do us part."

With all the empirical deviations, the love frame still reduces uncertainty in the negotiation of personal relationships considerably. Many of the activities connected to love are also expected between "best friends" and siblings: empathy, spending time together, supporting each other in times of need. But importantly, two realms are normatively barred from other personal relationships: building a household and family, and sexuality. Sexual intimacy (from kissing to intercourse) is often expected to occur only when the vocabulary of love is successfully emulated. As I mentioned, the notion of romantic love has now been extended to homosexual relationships, but the sharp distinction between love and non-love ties is mostly upheld. There are many circumstances that forbid ties to become "romantic love" and to involve sexual intimacy. We are not expected to become intimate with our siblings or with workplace superiors/subordinates. Also, people already engaged in love relationships are off-limits, at least in principle.

All of these rules are breached from time to time. Institutional prescriptions are never followed to the letter. Foucault argues that "perversions" including homosexuality form part of the discourse of sexuality, and that their status as "forbidden fruit" adds to their attraction ([1976] 1978, 48f). Sexuality outside the norms of romantic love—homosexuality, extramarital affairs, incest, "friends with benefits," polyamory—has to be negotiated carefully, and marked as exceptions by the people involved and/or by the social environment. One way of facilitating these kinds of aberrations is their organization at particular foci of activities, like brothels and swinger clubs, or more recently online forums and mobile apps (like Grindr or Tinder). These foci are symbolically demarcated from "normal family life," as a way of supporting the norm against deviation.

Overall, the love frame gives us effective orientation of what to expect of each other: the many circumstances under which sexual behavior is expected not to occur, how to construct personal relationships with bodily intimacy (and the prospects of building a household and family), what to do in these kinds of relationships. The norm simplifies these negotiations by providing an institutionalized script for interaction. Deviance is improbable and rarely achieved.

6.3. Network theory

A further aspect of love relationships "on the ground" is their embedding in networks of personal relationships. If romantic love is a relationship frame, and if gender is a social category, both of them will affect not just individual relationships, but also their patterning (see section 2.7 and chapter 4). What theoretical arguments have been advanced concerning these network patterns?

Love and gender are firmly inscribed in kinship networks (Parsons [1943] 1954; White 1963; Schweizer and White 1998). Love is supposed to run between one man and one woman, their relationships lead to children, and these again form couples with new children. This results in the familiar grid of kinship ties with people connected horizontally (to their spouse and to siblings) and vertically (to parents and children). Culturally variable rules govern the selection of mates, like the marriages between cross cousins prescribed or forbidden, or the marriage across kinship groups (Lévi-Strauss [1949] 1969; Bearman 1997). Following François Lorrain and Harrison White the positions of kinship categories like "uncle," "niece," or "grandmother" follow structural equivalence: nieces need not be connected to each other, but they occupy similar positions in the family web (in relation to other kinship categories (uncles, grandmothers, etc.); Lorrain and White 1971; Lorrain 1975; see sections 2.8 and 5.1). Things become more complicated with multiple marriages per person after spouses die or divorce, and with more openly homosexual relationships (sometimes involving children in various constellations). However, the grid of family relations still builds on heterosexual love and gender almost by definition.

In one of the first systematic network studies, British social anthropologist Elizabeth Bott examines the web of social relationships around married couples in mid–20th century London ([1957] 1971). She finds that the division of labor in the family heavily depends on its network embeddedness. The more husband and wife have close-knit networks around them, with many interpersonal ties to friends and family shared by both of them, the more they exhibit a gendered division of labor ([1957] 1971, 59). Looser-knit networks with a high degree of separation for husband and wife, in contrast, correspond to more sharing in household chores. Bott argues that these differences are related to class position: working-class families have more close-knit networks and therefore a more traditionally gendered social organization. More importantly, these differences result from spatial mobility: couples who move around have more acquaintances and friends who do not know each other, and this makes it harder for women to perform the household organization and child care normatively expected of them ([1957] 1971, 92ff). This also makes for a weakening of traditional gender norms, and for a stress on equality and joint family organization. Following Bott, we should not expect family relations to be the same across social contexts and social strata. But also, individual factors like mobility affect networks of personal relationships and family organization, as well as the rigidity of gender norms.

Cecilia Ridgeway, Lynn Smith-Lovin, and their coauthors adopt a wider perspective on gender. They see gender as a "master identity," a "frame" to categorize people in diverse contexts, leaving an imprint on them (Ridgeway and Smith-Lovin 1999, 205ff; Ridgeway 2009, 147f, 152). Ridgeway argues that the gender category projects differences in experiences with other people (1991; Ridgeway and Correll 2004, 517ff). These experiences get summed up over various situations into general expectations about people of a particular gender. Now these "gender beliefs" lead to particular patterns in behavior and to gender-typical network constellations (Smith-Lovin and McPherson 1993; Ridgeway and Smith-Lovin 1999, 193ff). Already among children, relationship ties tend to separate by gender. These "gendered networks" breed gender-specific belief systems, with gender-typical behavior ensuing. In adulthood, women are more enmeshed in dense familial and local networks, whereas men form more "weak ties" in the economic and public realm. These gender-typical networks reinforce the focus on relating within close-knit communities among women, and they make for more information and a broader perspective among men. Consequently, the gender distinction fosters the differences in behavior that it maps. The beliefs about gender differences become "self-fulfilling" (Ridgeway and Correll 2004, 518f, 523).

Ridgeway and Smith-Lovin do not confine gender to the realm of kinship relations. Rather, they account for the spreading of gender as an organizing principle beyond the family. Nevertheless, gender is not a free-floating category, but intertwined with the institutions of love, marriage, and family. The beliefs about gender formed there spread to other social contexts, including politics and the workplace. But the expectations connected to the relationship frame of love also have an impact on social networks beyond intimate relationships.

6.4. Sexual fields

The opposite strategy is taken by the recent formulations of a sexual field (Green 2014). True to the concept of field, relations of desire and sexuality are regarded as a separate realm of social life, disconnected from other social relationships like friendship and kinship. Following this approach, attraction and intimacy result from interaction and mutual observation in the field. Actors derive their attractiveness and desirability from displays of interest by others, leading to stratification in the field. Building on Bourdieu, this stratification is reflected in unequal distribution of "sexual capital" (Martin and George 2006). The positions in the field, and the effect of attributes like physical beauty or income on these, are negotiated in observation and interaction between the actors involved (Green 2011). Actors in the field are interested in sexual relations with people with more sexual capital, which would increase their own sexual capital. But they lose sexual capital if rejected. Therefore, they only approach others slightly higher up the ladder, or on the same bars, to keep the risk of rejection low. Only if the discrepancy of status in the field is not too high will these advances lead to sexual intercourse. The theory predicts what positions will initiate sexual contacts to what other positions in the field, and which of these will lead to sex.

Sexual field theory holds best for social contexts with two specific conditions:

(1) Everybody in the field is looking for sexual contacts, but not for *monogamous love relationships*. Whoever enters a traditional romantic love tie immediately becomes unavailable to others as a potential sexual partner, effectively leaving the field. Indeed, those looking for faithful, romantic love might not try to partner with actors high on sexual capital. Past success in the sexual field indicates a disposition for sexual promiscuity and proactivity and does not increase attractiveness for monogamous love ties.

(2) All actors in the field are available for sexual contacts with all others. Monogamy in intimate ties block this ideal market situation in one regard. Another block is *sexual orientation*. In a situation of all-around heterosexuality in a population 50 percent male and 50 percent female, only half of the dyads in the field can involve sexual contacts. We would have a field with two segments, where the actors compete within their own segment for attention from, and sexual contacts with, high-ranking members of the other segment. A heterosexual

field consists of two gender-pillars, with all relevant interaction taking place between them, and with all competition within them. The ideal of a social field, however, has all interaction and competition within one population of actors. The introduction of different categories of actors that interact according to specific rules constitutes an amendment to field theory, a deviation from ideal-typical conditions.

To save the theory, we can term both romantic love and the prevalent heterosexual orientation institutions that regulate the interaction in the field (see section 5.6). They prescribe the kind of interaction in the field, and between which actors it can occur. Adam Isaiah Green studied interaction in a field that lacks these institutions: the community of proactive homosexuals in Greenwich Village / New York City (2011). Here we find the hypothesized stratification of attractiveness out of mutual observation and sexual contacts in almost pure form. Other social contexts approximating the ideal of a sexual field would be polyamorous communities or swinger clubs, with widespread bisexuality and with romantic ties suspended, or not formed.

However, even in its pillarized heterosexual version, we could hypothesize to find the same mechanisms at play as in an unstructured sexual field (figure 6.1): Displays of interest are based on one's own sexual capital in relation to that of potential mates (with a preference for people slightly higher up the ladder). These are met with approval if the difference in sexual capital is not too big, leading to bodily intimacy. One's value in the field (sexual capital) would not directly be based on individual attributes, but on the interaction in the field, and on its observation by other actors. We can transfer these

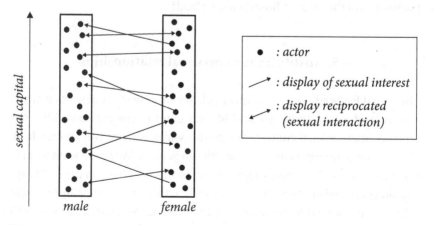

Figure 6.1. Patterns of interaction in a heterosexual field.

arguments to a world that is organized not around fleeting sexual contacts, but around relationships of romantic love.

However, both the heterosexual pillars and the longevity of romantic relationships make it hard to estimate our own sexual or romantic capital in relation to that of potential partners. Strictly speaking, in order to compare the value of an interested ego with that of a potential partner alter (of the opposite gender), ego would have to find people of the opposite gender with whom ego has already interacted sexually (whether only by display of interest or by sexual contact) and observe how they have interacted with people of ego's gender with whom alter has already interacted. Chances are that this constellation is quite rare, and there even seems to be a norm against it (Bearman, Moody, and Stovel 2004, 71ff). The longevity and monogamy of romantic love relationships (as opposed to fleeting sexual contacts) makes it even harder to compare one's value in relation to that of potential partners based on actual observations. But the actors in the field might develop a good sense of the worth of particular traits (education, power, beauty, charisma, age) that allows them to roughly estimate their relative sexual, or romantic, capital.

The network structures in a heterosexual romantic or sexual field would show a high degree of heterophily (with regard to gender) and marked homophily in a latent underlying status dimension (sexual capital). But this status dimension probably has limited explanatory power, given that men and women do not only partner based on perceived attractiveness to others in the field, but also on the basis of their age, education, interests, and styles. How much explanatory power does such an assumed latent status dimension need in order to count as a "structuring principle" constituting the "objective positions" (in the sense of Bourdieu) of a field?

6.5. Institutions in personal relationships

The sexual fields approach considers only sexual contacts and treats them as special (and as constituting a field of their own) in comparison with other kinds of interpersonal contact. Adopting a wider perspective, we can think of the realm of interpersonal relationships in general. We might want to term it a "field of personal relationships": actors in this field have an interest in rewarding personal interaction, from leisurely activities and empathy to bodily intimacy and sexual intercourse. But they face uncertainty as to who will engage with them in these kinds of personal interaction. This uncertainty is effectively reduced by adherence to relationship frames as institutions of the field. These define what kinds of interaction we can expect in what kinds of relationships, and whom to form them with. Love and sex are typical forms of relating in this larger field, as are friendship and family relations. If sexual contacts and love relationships are special, it is because we define them as special, relative to friends, siblings, parents, work colleagues, neighbors, etc. Sexual relations and romantic love are constituted by the cultural distinction between different kinds of ties and of activities in the first place.

Let me counterfactually consider a world of personal relationships unstructured by relationship frames (including love). In this world, we would engage in different kinds of activities—from shaking hands through going to the movies and empathic conversation, to kissing and sexual intercourse—based solely on how much we like another person, unhindered by institutionalized expectations about how to behave in particular kinds of relationships. We might end up kissing our friends and our siblings or even have sex with them. And we would have no need to call any of our significant others our partner or our spouse (and we would almost certainly not restrict bodily intimacy to one person). Now few of us would want to live in such a world or could even imagine doing so. This hardly points to the naturalness of our thinking and feeling in different kinds of relationships. Following Foucault, this thought experiment shows that even our desires are imprinted by cultural rules. In particular the twin institutions of romantic love and of gender have a thorough grip on our thoughts and our feelings, and on our ways of relating.

Returning to relational sociology, John Padgett sees family ties as constituting a separate network domain—interrelated with, but neatly separated from, other domains like business or politics (Padgett and Powell 2012b, 5f). However, we have to ask, what makes a relationship political, economic, or familial? As argued earlier, sexual, romantic, or family ties are not a natural species of personal relationships, nor are political or business ties. Rather, they become "special" through our symbolic separation of them from other interpersonal ties. Accordingly, we have to examine this symbolic separation—the cultural construction of particular practices (e.g., bodily intimacy, or financial credit)—as forming part of particular kinds of relationships, and the careful management of social relationships in light of such institutionalized expectations (Zelizer 2005).

In this perspective, *families* are not coherent entities, "relational subjects" (Donati and Archer 2015), or "concrete things" (Bunge 1996, 17). First of all, "family" is a way to construct some personal relationships as different from others, and more binding than them: a relational institution that brings order into our messy interpersonal networks. Blood is thicker than water. The resulting family networks are symbolically elevated relative to friendships and other interpersonal ties. Their prescribed rigidity and commitment make families "real" social configurations, to the extent that communication defines them as real and adheres to the institutionalized expectations linked to them. However, this need not be the case. Family ties do not have to be more important than those to friends and colleagues. The recent trend toward serial monogamy and the resulting patchwork families have put the "natural" boundaries of the family under pressure, with institutionalized expectations eroding. Unlike formal organizations, families do not have paid contracts to demarcate their personnel and to ensure its commitment. However, they are certainly fraught with economic and legal implications (Zelizer 2005).

To summarize, gender and love structure personal relationships as dominant institutions, ordering symbolically which kinds of ties are permissible and which ones forbidden, or at least, improbable. As long as we think of (romantic) love and family as natural ways of ordering our relationships to others, the gender distinction is also involved. Consequently, one prime location for the construction and legitimation of gender differences lies in the relationship frames of love and family relations. Women and men are assigned the roles in social life compatible with our understanding of love. As this understanding changes, so do the expectations tied to the genders.

Combining relational sociology and poststructuralism, Lotus Seeley similarly conceptualizes gender not as an individual trait, but as a culturally variable category inscribed in network patterns (2014, 37ff). Therefore, we have to examine the stories woven around the gender category and the network patterns stabilized by it. According to Charles Tilly, such stories make sense of, and legitimize, categorical distinctions, and they effectively constrain our interaction (1998, 63f, 102f). Expanding on the formulations of Tilly and White, we have to pay attention to the "private stories" of love and gender in network domains, *but also* to the "public stories" diffused in the mass media and in systems of education. The institutions of love and gender are not only reproduced on the level of relational expectations in interpersonal ties (the "meaning structure of social networks" from section 2.1).

They are themselves cultural constructions and part of the wider cultural environment—generalized from concrete relationships, and amenable to representations in the mass media, for example, in novels and Hollywood movies (see sections 2.2 and 5.9).

6.6. Structural equivalence

In line with these arguments, gender is conducive to *structural equivalence*, not only in family relations, but in networks of personal relationships generally. Lynn Smith-Lovin and Miller McPherson find empirical evidence of the structural equivalence of men and women, respectively, in their embeddedness in networks (1993, 228f). The concept of structural equivalence comes with a number of implications (see sections 2.8.c and 5.1):

- Structurally equivalent actors occupy *similar positions* in networks, without necessarily forming cohesive groups. Men and women interact too much and too intimately with each other in order to form densely connected (and structurally separated) interaction groups like ethnic categories, for example. Rather, the institutions of gender and love make for the particular positioning of men and women, respectively, in social structure.
- As in blockmodel analysis (White, Boorman, and Breiger 1976), this positioning pertains to *different types of tie*. Here we would expect love and gender to govern romantic love, but also kinship and friendship ties.
- In the original formulations, structural equivalence and blockmodel analysis were connected to the concept of *roles*. Roles denote typical bundles of relationships tied to a particular *position*. Roles and positions can emerge endogenously, or they can be imported from formal organization or cultural discourse. In both instances, these serve as recognizable patterns that actors orient to and adhere to (see chapter 5).
- Roles and positions are inferred *inductively* by way of formal analysis of relationship patterns, rather than derived from discourse about social structures (in line with Emirbayer and Goodwin's "anti-categorical imperative"; 1994, 1414).

Some of these arguments change when applied to gender. We know the gender category and the relationship frame of love to be important in social life, and we hypothesize that they lead to patterns of structural equivalence. Therefore, we cannot follow the anti-categorical imperative too strictly. Disregarding categories sounds like a good idea theoretically and ethically. But it is not practical if we want to know about the consequences of categories in the social world. These have to be explored and tested, rather than assumed or dismissed.

6.7. Network patterns

Combining the concept of structural equivalence with these arguments yields a number of theoretical expectations about the network embeddedness of romantic love and gender, in their prototypical modern installments:

> Conjecture 6.1. Personal relationships are ordered by gender, with love ties predominantly heterosexual and friendships mainly to the same gender.

That love ties should be heterosexual seems to be true by definition. The institution of romantic love is still, by and large, defined as running between one man and one woman. This heterosexual orientation is gradually eroding over the last decades, but it mostly holds for now. In the GSS, respondents were asked for their sexual orientation since 2008. In spite of a marked increase of people indicating homosexual or bisexual orientations, over 92 percent of respondents in 2016 report being "heterosexual or straight."

If the traditional blueprint of romantic love is heterosexual, friendship tends to run within gender. We know that a relationship between a man and a woman can turn into romantic love (unless they are defined as ineligible, for example, as siblings). Their dealings with each other will be complicated by this possibility. If A and B get along well with each other, why not become intimate? The framing of male–female relationships as "friendship" reduces the inherent uncertainty of all interpersonal communication only partly. Cross-gender friendships should be more fraught with misunderstandings or even with asymmetric romantic interest, and less stable than male–male or female–female friendships—as long as those are not seen as potentially leading to love.

But male–female friendships are not only less stable inside. They are also suspicious to outside observers for their romantic/erotic possibilities. A cross-gender friendship between A and B will be frowned on if

A is already involved in a romantic relationship with somebody else. Her partner (C) might doubt A's fidelity, or her motivations for befriending B. Consequently, A might withdraw from her friendship with B in order to reduce tensions in her love relationships. These are just potential factors making friendships within gender more likely than cross-gender. Probably, people do not really ponder these when befriending somebody (or not). As a consequence of these problems, the cultural *model* (the relational institution) favors friendships to the same genders. The predominance of same-gender friendships follows the script rather than results from actual problems.

Finally, we could argue that people of the same gender share similar experiences and a similar perspective, due to their structurally equivalent positions (Erickson 1988, 109ff; see section 2.8.c). This makes friendships as a type of intimate relationship focused on mutual understanding and shared activities likely. The same argument should also apply to romantic love. But here the motivation for sexual interaction (culturally defined and subjectively desired as cross-gender; Foucault [1976] 1978) seems more important, even at the expense of empathy and interests.

Empirical studies show a high propensity for same-gender friendships among adults (Verbrugge 1977, 586f). Especially marked is the tendency for same-gender friendship choices among schoolchildren, only mildly varying through the grades (Shrum, Cheek, and Hunter 1988). Sanne Smith, Ineke Maas, and Frank van Tubergen find homophily by gender to be stronger and more consistently predictive for friendship nominations in schools in four European countries than homophily by ethnicity, religion, socioeconomic status, and risk behavior (2014). Therefore, boys' and girls' networks are sometimes even analyzed separately for meaningful results (Lubbers 2003). In contrast, women and men have increased in their tendency to name confidants from the opposite gender in the GSS from 1985 to 2004 (Smith, McPherson, and Smith-Lovin 2014, 440, 443f). This change is mainly driven by the increase in the naming of spouses as (cross-gender) confidants. The tendency to "discuss important matters" with the same gender, outside of love relationships and the family, holds by and large steady.

Conjecture 6.2. Love and friendship lead to particular network patterns. Love is dyadic and exclusive, whereas friendship tends to transitivity.

Relationship frames like "romantic love," "friendship," and "patronage" do not only prescribe what kinds of people to relate to in these ways, and what kinds of behavior these kinds of relationships allow for (or even demand). They also come with prescriptions regarding the network patterns of these relationships. Most obviously, relationships that adopt the "romantic love" frame require the partners to forego other romantic entanglements, or any other relationships involving bodily intimacy (such as "affairs" or "friends with benefits"). Consequently, these relationships are mostly isolated dyads, unconnected to other romantic relationships or sexual contacts.

Are romantic relationships really that exclusive? Based on recent surveys, about 2–4 percent of spouses in the United States engage in extramarital sexual contacts in any given year, and infidelity ever occurs only in 20–25 percent of all marriages (Fincham and May 2017). Most sexual contact outside of romantic relationships remains episodic, rather than developing into longer affairs. The basic norm of having only one intimate *relationship* at any point mostly holds. Looking at sexual contacts among adolescents at one U.S. high school, Peter Bearman, James Moody, and Katherine Stovel find "a large number of nonromantic sexual partnerships" (2004, 57). This is in line with changing attitudes on sexuality and love relationships, and with a widespread tolerance for sexuality outside of romantic love among young people in the United States (Friedland et al. 2014). However, more than one-fifth of the high-school students (126 out of 573) are only connected in isolated heterosexual dyads over the last 18 months (Bearman, Moody, and Stovel 2004, 59). Many more report only one sexual relationship, but their partners had sexual contacts with others. Even among adolescents, the dyadic romantic tie continues to act as a model that most sexual contact conforms to.

If romantic love is exclusive and dyadic, friendship tends to be inclusive and transitive. Following the logics of balance theory, I treat my friends' friends amicably (Cartwright and Harary 1956). First, I might trust my friends' taste and assume that I will get along well with their friends. Probably, this is due to the fact that we often form friendships with people with similar interests and attitudes (Erickson 1988). Chances are high that I am closer in interests and attitudes (in cultural imprint; see section 2.8) with my friends' friends than with random acquaintances. Second, I have a higher probability of encountering, and getting to know, people through a common friend. This increases the odds of forming a friendship with them (but also of developing a love tie). Third, it would strain our relationships with our common friend if we did not get along with each other. Befriending each other results in the balanced configuration of a closed triad. However, as the example of love shows, this tendency toward transitivity in positive personal relationships is not a quasi-natural structural force but depends on the culturally variable expectations we connect to different kinds of relationships. A triadic constellation with two friendships tends to balance by the formation of a third friendship (closure). But in a triadic constellation with two love relationships, one of these is normatively required to dissolve, at least in contemporary Western culture (see section 2.7).

Studies of friendship networks in school classes show a marked tendency toward transitive closure of friendships (e.g., Lubbers 2003, 322ff). Transitivity leads to the widespread formation of cliques, all of which develop their own group cultures and identification (see chapter 3). As John Levi Martin argues, transitivity and cliques occur most often in social contexts where the same sets of people interact frequently and for long periods of time (2009, 26ff). Such "caging" contexts, like schools or prisons, are precisely those lending themselves most conveniently to studies of full networks. It is relatively easy to establish the boundaries of such networks, and to get most people in the network to fill in questionnaires. Outside of schools, prisons, and workplaces (and of international politics), we find less transitivity. If I don't get along with my friends' friends, I can just avoid them—unless we are all in the same school, in the same prison, or working in the same office.

In comparison with the typical clique structures of school friendships, the network of sexual contacts reconstructed by Bearman et al. is striking for its lack of transitivity. The network of 573 students features no 3-cycle of three actors all connected to each other (2004, 58). This can be explained by the almost complete lack of homosexual or bisexual contacts.[4] The institution of gender alone would then account for the tendency against closure. But Bearman and his coauthors also find almost no 4-cycles, though we would expect them to occur frequently if partner choice were solely based on gender and on similarity of style, attitudes, and attractiveness (2004, 66, 71ff). They explain this through a norm against "dating your ex-partner's ex." As a result, the network of sexual contacts forms the structure of a spanning tree, with paths of different length branching out and bifurcating into sub-branches. This looks very different from the gendered friendship cliques in school networks.

---

[4] For transitive closure (3-cycles) to occur in sexual contacts, we would need either two bisexual actors connected to each other homosexually and to a third actor of the opposite gender, or three actors of one gender connected through three homosexual ties.

Conjecture 6.3. Women typically maintain stronger ties to children and to the wider family. Men have more ties with colleagues and in the economic and political public.

To the extent that the division of labor in romantic love and in the bourgeois family household remains gendered, we expect men and women to connect to different social realms. Women would be responsible for child care and for family-related ties. If their culturally assigned place is "in the family," the family is also held together by them. Men, in contrast, would be freed from extensive household chores, including child care. Their primary realm of interaction is the world of business, paid work, and politics. This division of labor is surely changing. Women are advancing into the labor force and into public-political relations, and men are increasingly involved in the upbringing of children and in the household management. In any case, both the institution of the gendered bourgeois household, and its change, should be reflected in the network embeddedness of men and women.

Empirical studies show women to name more kin, and men more non-kin, as close associates (Fischer and Oliker 1983; Marsden 1987, 126f, 129; Moore 1990). But these tendencies vary over the life cycle, with young people socializing more with non-kin, and with older respondents turning inward to the family for close associates. To my knowledge, gender differences in ties to work colleagues, in business, and in politics have not been examined. Matthew Brashears finds relatively little difference in the tendencies of women and men to associate with confidants with a similar level of education, of a similar age, and from the same religion (2008, 2015).

John Padgett and his collaborators investigate political, business, and marriage ties between patrician *families* in Renaissance Florence, rather than between individuals (Padgett and Ansell 1993; Padgett and McLean 2006; Gondal and McLean 2013). Most of the business and political ties were formed by men, with women mainly involved in the marriage ties. This research justifiably builds on the idea of the household as a unit, with men as representing the household in politics and in business. But with the gradual incorporation of women in the world of paid work, we want to know to what extent their networks change (and those of men).

We do have extensive studies on networks in companies, some of them focusing on the role of gender. Herminia Ibarra finds that men and women, respectively, form network ties mainly among themselves (1992). But women have more cross-gender ties than men. No big surprise if the company is numerically dominated by men. The latter are also more central in all types of networks considered (communication, advice, support, influence, friendship). This supports the "old boys' network" hypothesis. Women have a hard time breaking into the informal ranks of men in business, and this probably contributes to their dim chances of advancing formally. Ronald Burt argues that the mechanisms for upward mobility of managers in a large American company differ by gender (1998). His notion of "structural holes" based on "weak ties" seems to have less explanatory power for the professional success of women. He goes on to identify a different pathway of success for women in their relationships to superiors. According to Burt, women act as an outsider group in the manager networks and therefore have to "borrow" social capital from powerful male patrons.

Family and workplace are foci of activity in the sense of Scott Feld (1981). These provide the opportunities for interpersonal communication that can lead to personal relationships (see section 2.1). If women spend more time at home, and men more time at work-related settings, this almost inevitably leads to gendered personal networks. However, these foci of activity are not mere structures but are imbued with cultural meaning (see section 5.10). Here, the expectation that the place of women is at home leads to the gendered composition at the workplace, but also to better chances of advancement for men.

Combining these arguments, we can depict the embeddedness in networks expected for the genders as in figure 6.2. Women should have stronger ties to their children and to other kin than men. And they should mostly have female friends. Men maintain weaker ties with their children and with the wider family. Most of their friends should be male. Some of these friendships

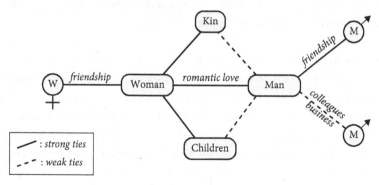

Figure 6.2. Gender-typical network embeddedness.

will be formed in the world of work and business, where they also keep a lot of weak ties to colleagues and business contacts. At the center of this gender-typical embeddedness, though, lies the heterosexual tie of romantic love. Without it, friendships could be more often cross-gender, and women would not be tied to their family (but would pursue business careers and form more ties there).

In a sense, this figure captures only "what everybody knows" about the world of personal relationships. I do not see this as a weakness. Actors have to know the cultural rules of relationship formation to follow them. However, this familiarity with the patterns of intimate and other personal relationships may give us a false sense of confidence in these theoretical expectations. My aim here is to systematize our knowledge about love and gender and their role in personal relationships. Doing so will hopefully get us beyond "what everybody knows" and allow us to distill some nontrivial theoretical expectations, but also to confront this more systematic framework with empirical evidence.

6.8. Gendered confiding relationships

As a small and very limited test of these considerations, I have a brief look at how the genders differ with regard to their close confidants. The data is from the *GSS* 2004, the last time the extended module on egocentric networks was included. Respondent were asked the following question ("Burt name generator"), and to give information on up to five alters:

> From time to time, most people discuss important matters with other people. Looking back over the last 6 months, who are the people with whom you discussed matters important to you?

We know from past research that people tend to not focus on actual discussion of "important matters" in their answers. Instead, they indicate the most important people they typically talk to about a variety of topics—their confidants (Bailey and Marsden 1999; Bearman and Parigi 2004; Small 2017). The number of confidants tends to be low—between two and three. We can assume that the respondents named people with whom they had *strong ties*, with a focus on empathy, irrespective of whether these relationships also involved other social activities like bodily intimacy or material support. In 2004, 1,070 people were asked and named at least one confidant; 356 did not name any. Miller McPherson, Lynn Smith-Lovin, and Matthew Brashears interpret this as indicating social isolation (2006). But it seems that many people simply chose not to give information on their confiding relations, sometimes advised by their interviewers (Fischer 2009; Paik and Sanchagrin 2013).

In the following, I compare the 1,070 respondents who reported at least one confidant. On average, women reported more people they discussed important matters with (Table 6.1). Therefore, the following analyses will be slightly biased by gender. But the respondents also named more women as confidants—and this holds true if we control for the tendency of women to name more confidants. Probably, women simply have a slightly larger number of confidants, but not to the extent indicated by the raw numbers. However, men reported more talking to male confidants (1.44 vs. 1.14), and women indicated more female discussion partners (1.67 vs. 1.04). There is some homophily by gender, but we find only a modest bivariate correlation between the genders of the respondents and of their confidants (Pearsons $r = 0.142, p < 0.001$). In comparison with the GSS data from 1985, the gender homophily of confidants is declining (Smith, McPherson, and Smith-Lovin 2014; see section 6.7).

Table 6.1. Confidants in the GSS by gender (arithmetic means)

	Female Confidants	Male Confidants	Sum
Women (612)	1.67	1.04	2.71
Men (458)	1.14	1.44	2.58
All respondents (1070)	1.44	1.21	2.65

We do not have to find a prevalence of within-gender ties to detect structural equivalence. Rather, men and women should diverge in the kinds of relationships (types of tie) to confidants. In the following, I analyze whether male and female respondents systematically differ in whether they name people from various categories as confidants. These categories include "spouses," "friends," and siblings, indicating kinds of relationships rather than kinds of people. For these theoretical expectations to hold, men would have to name more male friends and colleagues. Women, in contrast, should report more kin and children, and more female friends. Since only strong ties are considered, the hypothesized weaker ties of men to children and to other kin should not feature to the same extent.

Given the structure of the data, I binarized for respondents of a particular category named. Only a few cases have more than one confidant per category, for example, if somebody reports discussing important matters with both parents. A binary logistic regression then tests whether men or women are more likely to name at least one confidant of the various kinds of alters. I control for level education, age, place of residence (rural/urban), work status (working part-time, temporarily not working, unemployed, housekeeping, attending school, with working full-time as reference category), marital status (married, widowed, divorced, and separated, with never married as reference category), and number of alters named in the interview. I also test whether one of the genders has a higher density in its egocentric networks. Density is calculated by dividing the number of alter–alter relationships reported by ego through the possible ones in the respective egocentric network. This can only be measured and analyzed for the 786 respondents who named at least two confidants. Because of its highly discrete distribution, I also binarize density into above 50 percent (dense) and up to 50 percent (loose), with slightly more respondents (56.5 percent) reporting dense egocentric networks.

Figure 6.3 presents the odds ratios of naming at least one confidant from each category by gender. An odds ratio of 1 means that both genders are equally likely to name a particular category of confidant. Therefore, 1, rather than 0, is the neutral line in the middle. If men are more likely to name a kind of alter (as listed on the left side, for example: coworker), the bar for the effect goes to the left. Bars to the right mean that women are more likely to name these alters as confidants, controlling for education, age, rural/urban residence, work status, and marital status.[5] The control variables with a significant effect ($p < 0.05$ in a two-tailed test) are listed on the right side of the figure. Unsurprisingly, the number of alters significantly increases the odds of naming any one kind of confidant (and, counterintuitively, the density of personal networks). This trivial effect is not listed in the figure.

---

[5] In line with the statistical model of the logistic regressions, the scale of the effects is logarithmic in the diagram. The partial effects are listed as odds ratios to facilitate interpretation.

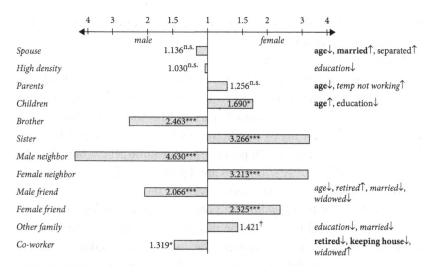

Figure 6.3. Relative odds of naming confidant in *GSS* 2004 by respondent gender.

*** $p < 0.001$, ** $p < 0.01$, * $p < 0.05$, † $p < 0.1$, n.s. = not significant ($p > 0.1$); two-tailed tests.

Additional effects are listed if $p < 0.05$. ↑ signifies an increase, and ↓ a decrease in the odds of naming the kind of confidant.

Control effects in bold: $p > 0.001$, normal: $p < 0.01$, italics: $p < 0.05$ (two-tailed tests).

As the figure shows, men slightly, but not significantly, more often name their *spouses* as confidants (+13.6%). Older respondents tend to confide less, and married and separated respondents to confide more in their spouses. There is virtually no effect of gender on network *density*, but more educated respondents tend to have looser networks (Lizardo 2006). In line with the theoretical expectations, more women than men discuss important matters with *parents* (+25.6%) and *children* (+69%, $p < 0.05$), even if the effect for parents is quite uncertain. Unsurprisingly, older respondents name parents less often, and children more often.

Members of the *wider family* are more often confided in by women (+42.1%, $p < 0.1$). More notably, the probabilities of discussing important matters with siblings differ dramatically by gender: *sisters* are more than 3 times as likely to be named by women, and men name their *brothers* 2.5 times as often (both effects: $p < 0.001$). Male and female respondents are equally likely to have brothers or sisters. And cross-gender siblings are morally ineligible for romantic relationships. Therefore, this effect probably reflects gender-typical bonding. People might confide in their same-gender siblings more because they share similar experiences and worldviews, due to their gender-specific positions in the social world.

Neighbors of the same gender are confided in 4.6 times (men) and 3.2 times (women) as often as those of the opposite gender. The same holds for "*just friends*," but with slightly lower odds ratios (men: 2.1×, with significant control effects for age, being retired, married, and widowed; women: 2.3×). All of these effects are highly significant ($p < 0.001$). Women and men are a lot more likely to socialize with people from the same gender outside of the family. Given that mostly strong ties are reported, neighbors probably also count as friends—significant alters outside of the institutionalized family roles (Fischer 1982b). Neighbors and other friends are preferred from the same gender as confidants, meaning that they share similar gender-specific positions, perspectives, and interests. Also, same-gender confidants are usually not seen as candidates for love relationships, or as threatening existing ones. This makes it easier, and more permissible, to confide in them. The findings pertaining to neighbors are less noteworthy because of their magnitude. Relatively few respondents report discussing important matters with their neighbors (117 vs. 722 respondents for friends). This renders the estimated odds ratios less confident. But unlike many other friends we make at work, doing sports, or in university, we meet male and female neighbors in the same social contexts, unstructured by gender or by gender-specific interests or occupations. Here opportunity for contact is controlled for, but we are far more likely to make friends and discuss important matters with the same gender, at least outside of love relationships and the family.

Finally, men are significantly more likely to report discussing important matters with *colleagues from work* (+31.9%, $p < 0.05$). This conforms to the institutionalized expectation that men are assigned the role of main breadwinner and responsible for public contact of the family. This effect is not specific to alters of different genders.[6] Rather, the point is to show the overall effect of gender on access to the world of work through strong ties. This overall effect is weaker than that to friends, neighbors, and siblings by gender. But it holds after *controlling for employment*. If women work part-time and keep the house more often, these effects have to be added to the already significant direct effect of gender.

---

[6] Women are 2.49 times as likely to confide in a female coworker, and men 5.08 times as likely to confide in a male coworker.

Women tend to have their confidants in the family, and to add female friends, some of whom they meet in the neighborhood. While women focus more on the family for socializing, men are more likely to make friends at work. This allows them to discuss work-related issues more. Consequently, their world centers less on children and on family issues, and more on jobs and careers. Of course, this difference partly rests on opportunities for contact. Trivially, people keeping the house (mostly women) less often discuss important matters with colleagues from work. But the effects reported here hold in addition to these structural opportunities, as captured by the control variables. Gender affects interpersonal relationships as a normative model for how to relate to whom over and above structural factors.

6.9. Conclusion

In this chapter, I have developed a relational perspective on love and gender. On the one hand, this locates love and gender in a universe of symbolic relations, similar to the one reconstructed by Bourdieu among the Kabyle. In this symbolic universe, the gender category is tied to the cultural model of heterosexual romantic love. And romantic love is firmly embedded in the cultural repertoire of family relations but also bears similarities (and important differences) with other *relationship frames* like "friendship" or "love affair" (Fuhse 2013; see section 2.7). All of these cultural models come with prescriptions for relational expectations, and for the kinds of communication, between the actors involved. But they also imply rules for how to connect to third parties.

Love and gender are inscribed in the meaning structure of social networks, making for particular network compositions and network patterns (and for gendered patterns of communication). Romantic love demands exclusivity. Love relationships are mostly confined to isolated dyads, the vast majority of them heterosexual. But these love dyads mesh in wider webs of friendship and kinship relations. In particular, romantic partners are incorporated in each other's family. And they tend to produce and raise children. Family relationships are organized around romantic love (Ketokivi 2012). In its traditional version, romantic love assigns the tasks of child care and of tending to kinship relations to women. Men, in contrast, are better connected to the worlds of work, business, and public-political relations. This partly stems from their structural access to job markets and to the political realm.

However, these foci of activity are themselves organized around cultural notions of who should have (what kind of) access to them, and who should advance to better jobs or in political careers.

Love and gender, then, can be termed *relational institutions*: cultural models for relationship formation and for network patterns (see chapter 5). Like all institutions, love and gender are models adhered to imperfectly, and they vary over time and across social contexts. But in any given context, the institutionalized expectations tied to these models should lead to the *structural equivalence* of men and women, respectively. Structural equivalence holds across different types of tie. For example, love relationships are still mostly heterosexual. Friendships tend to form within the gender categories. The empirical analysis of data from the *GSS 2004* shows that women also have more strong ties to children and to the wider family (section 6.8). Brothers frequently act as confidants for each other. As do sisters. Brother and sister confide in each other less often, making sibling relationships strikingly gendered. Men also name more colleagues from work as their confidants, even after controlling for employment.

The expectations tied to the gender category and to the relationship frame of romantic love continuously change, albeit slowly. The traditional gendered version of romantic love is currently making way for more "androgynous" (Cancian) ways of constructing intimate relationships. And sexuality is less confined to the relationships emulating the traditional ideal. Unconventional arrangements like "flings" or "friends with benefits" are becoming more acceptable and widespread, especially among young urbanites. Finally, homosexuality has made its way from the hidden dark corners of society to open and legitimate relationships, with objections becoming rare. Most people (and public stories) now accept that romantic love is possible between members of the same gender, and just as "special" and dependable as that between men and women. All of this should slowly make for the gradual decoupling of the gender category from the notion of romantic love (and family), and for a loosening grip of institutionalized models on intimate life and personal relationships. And it should lead to increased uncertainty, if established scripts for bonding no longer hold. Before this erosion of institutionalized expectations, it was easier to determine how people stood toward each other, or what kinds of relationships they could develop.

These trends should not be overstated, though. The empirical analyses show that personal relationships in the United States are still widely governed by gender and by relatively traditional ideals of romantic love, friendship, and family. Future analyses could focus on changes over time, and at disparities between more traditional and more progressive milieus (Bott [1957] 1971).

Finally, I briefly contrast this approach to love and gender to other theoretical approaches.

Peter Blau sees love, and other intimate relationships, as developing in the course of exchanges (1964, 76ff). These exchanges bring gratifications for the people involved, and they lead to obligations between them. I agree that social relationships develop in the course of communicative events, whether we term them exchange, interaction, transactions, or communication (chapter 7), and not out of preestablished individual dispositions. But Blau's model disregards the role of institutionalized scripts—relationship frames. These tell us with whom we can build what kinds of relationships, and what activities are expected in these relationships. Otherwise, we would not understand the strict boundaries between love, friendship, and family ties. These relationship frames, and romantic love in particular, also partly account for the asymmetry of expectations between partners. Blau's model explains such asymmetries structurally from the differential involvement in other relationships, ignoring the cultural prescriptions of the gender category.

My approach also runs counter to macro-sociological explanations of gender inequality out of the social organization of violence and economic markets (e.g., Collins 1971, 1991; Elias 1987; Chafetz 1990). Changes in the occupational structure and in the role of violence and of the military in society undoubtedly play a role for the status of the genders. But I claim that we first have to examine where and how the distinction between men and women is constructed, and how these two categories relate to each other. This "relating" of men and women takes place both on the cultural level of institutions, and "on the ground" in actual personal relationships of friendship, love, and family. This is not to dismiss the significance and importance of other factors. For example, we could think of a process through which particular professions typified as female become vastly important in social and economic life, or of another process with women successfully taking over key positions in the political elite. Both processes would lead to an increase in symbolic prestige of women, and to a change of expectations tied to the category. My conjecture is that these symbolic shifts would affect not only the gender category itself, but also the relationship frames of love and family.

Constructing an encompassing theory of gender stratification is a burdensome task that does not lend itself to simple models (Collins et al. 1993). I believe that the systematic symbolic connection between our constructions of gender and of romantic love gives us an important key to these phenomena. Theoretically, this follows from the assumption that forms of meaning make for particular ways of relating and are tied to them. Therefore, the gender category connects with the relationships between men and women. Gender relations are not just abstract symbolic, or economic, or political. They are also, and perhaps more fundamentally, grounded in the relationships we construct between and within the genders. Our ideas of love and gender have an impact on the patterns of personal relationships, as well as depending on and being reproduced in them.

7

Events in Networks

The last chapters have dealt with the interplay of social networks with forms of meaning. These two layers are seen as interwoven, based on the general stance of relational sociology (Pachucki and Breiger 2010; Mische 2011; Fuhse 2015b; McLean 2017). This amends and corrects the simplistic structuralist model of social networks as the mere pattern of links between nodes, or ties between actors (see chapter 2). Culture is no longer conceived of as the passive result of network formation. Instead, social networks are themselves viewed as imbued with meaning and inscribed with culture. They consist of relational expectations that are subject to cultural imprint: they build on relationship frames like love, friendship, or patronage (see section 2.7), and they are patterned by social categories like ethnicity (chapter 4) and gender (chapter 6) and by more or less institutionalized roles (chapter 5). All of these cultural imprints are realized and negotiated between the actors involved, but these negotiations draw on cultural models for how to relate—institutions (chapter 5).

This turn toward culture and meaning leads to a less structuralist view on social networks, and to important insights on sociocultural constellations. However, it leaves the static vision of social constellations inherent in network research intact. Both networks and culture are thought of as stable patterns that differ between contexts. Sometimes, networks are even compared across time, with some ties forming, and others disappearing (Moody, McFarland, and Bender-deMoll 2005; Snijders, van de Bunt, and Steglich 2010). The implication is: at any point in time, we can distinguish between existing ties and nonexistent ones. Of course, we can still distinguish between different types of tie, and between ties of varying strength, recognizing that social relationships carry different meanings and importance. But all of this relies on a unitary conception of social ties as things that are there, or not, and stable over time—unless they mysteriously spring into existence or vanish.

This still holds for recent advances dealing with the *dynamics* in network constellations (Moody 2009; Snijders 2011). These advances focus on

Social Networks of Meaning and Communication. Jan Fuhse, Oxford University Press. © Oxford University Press 2022.
DOI: 10.1093/oso/9780190275433.003.0007

the appearance or disappearance of ties over time. They mostly model network constellations formally, with ties switching from 0 (no tie) to 1 (tie), or vice versa. The same applies to the well-established network mechanisms like transitivity and homophily that are seen as "causing" observed network constellations (Monge and Contractor 2003; Rivera, Soderstrom, and Uzzi 2010). These models remain silent about the micro-processes in relationships—they are formal models, not substantive theories that would render the mechanisms at play plausible.

In this and the following chapter, I argue against the conception of ties and networks as stable and unproblematic patterns. The social world does not, at the very bottom, consist of social relationships, not even if we allow them to appear and disappear over time. Rather, it is composed of ephemeral events that vanish as soon as they occurred (Abbott 2016). In this sense, we have to think of social relationships as series of events, or as patterns in these series (Hinde 1976). How do these series stabilize (if and to the extent that they do)? Why do we find the kinds of regularities in ephemeral events that make observers and participants think of relatively stable relationships, like love, friendship, alliance, conflict, or patronage? Events seem to stabilize into relationships and networks, and these, in turn, govern future events. The notions of relationships and networks become problematic here, but they do not dissolve. Instead, we have to reconstruct what these concepts mean, in a social world fundamentally composed of events.

Returning to the framework from section 2.1, the last chapters dealt extensively with the interplay between the meaning structure of social networks (its relational expectations) and the wider cultural context. The turn to events now moves to the "patterns of communication" as the other side of social networks. So far, I have treated communication only in passing, as something that results from opportunities for contact and that somehow leads to relational expectations. But how exactly does this happen—where do relational expectations come from (as opposed to factual, temporal, and other kinds of expectations)? Now the processes in networks take center stage—the communicative events making for the emergence, stabilization, and change of networked expectations. We first have to find a convincing concept for these *events*. They are here considered as transitory occurrences that lead to future occurrences. Therefore, the connections between events should be part of their conceptualization—and in this sense, series of events are *processes* in social structures.

I start this chapter with a short overview of how the main authors of relational sociology think of relationships and processes (section 7.1). Then, I discuss seven prominent concepts for social events with regard to their implications and their usefulness for a theory of social networks: behavior (section 7.2), action (section 7.3), practice (section 7.4), exchange (section 7.5), interaction (section 7.6), transaction (section 7.7), and switchings (section 7.8). From this discussion, I distill a number of conceptual desiderata (section 7.9). The conclusion points out that Niklas Luhmann's concept of communication (the eighth concept) meets the desiderata quite well (section 7.10). The following chapter (chapter 8) develops this option in detail, laying out the communication theoretical perspective on social networks.

7.1. Processes and relations

As elaborated in chapter 2, relational sociology around Harrison White incorporated a concern for meaning in the 1990s. From the end of the 1990 onward, the processes in networks were added as a second focus. White himself develops the notion of "switchings" between network domains as foundational for change in sociocultural constellations (1995a; Mische and White 1998). Mustafa Emirbayer argues in his "Manifesto for a Relational Sociology" (1997) that relational sociology builds on "trans-actions" (building on Dewey and Bentley) transforming the actors involved (1997, 286f). Patrick Thaddeus Jackson and Daniel Nexon proclaim a "processual relationalism" for the study of world politics (1999). Around 2000 and shortly afterward, the empirical studies of Paul McLean (1998, 2007), Daniel McFarland (2001, 2004), Sophie Mützel (2002), David Gibson (2003, 2005), and Ann Mische (2003, 2008) cumulated to a thorough research focus on events in networks. This connects with authors from outside relational sociology who investigate processes of communication in networks (Butts 2008; Kossinets and Watts 2009; de Nooy 2011; Kitts et al. 2017).

If communicative events are now firmly established as a research focus, their theoretical conceptualization and that of the interplay with relationships and networks remain fragmented. The following quotes illustrate both the fundamental insight that networks and ties consist of events among relational sociologists, and the variety of theoretical figures for this:

the transactional approach [...] sees relations between terms or units as preeminently dynamic in nature, as unfolding, ongoing processes rather than as static ties among inert substances. (Emirbayer 1997, 289)
ties are not static "things," but ongoing processes. (Jackson and Nexon 1999, 292)

we should not see networks merely as sites for or conduits of cultural forms, but rather we should look at how both of these are generated in social practices, that is, by the dynamics of communicative interaction. (Mische 2003, 262)

interpersonal transactions compound into identities, create and transform social boundaries, and accumulate into durable social ties. [...] Strictly speaking, we observe transactions, not relations. [...] From a series of transactions we infer a *relation* between the sites: a friendship, a rivalry, an alliance, or something else. (Tilly 2005a, 7, original emphasis)

Any "structure" refers simply to specific transactions that share some similarities and that are more or less reproduced through time and space. (Dépelteau 2008, 69)

we begin by considering social structure simply as regular patterns of interaction, and leave to the side the question of why these patterns exist. (Martin 2009, 7)

the actual substance of any relationship consists of the specific interaction that goes on between the interconnected parties (Azarian 2010, 326).

Social relations, I want to suggest, are *lived trajectories of iterated interaction*. [...] A social relation is not an object, akin to a bridge, but rather a shifting state of play within a process of social interaction. *To say that two actors are related is to say that they have a history of past and expectation of future interaction and that this shapes their current interactions.* (Crossley 2011, 28, original emphasis)

Understanding relations themselves as processes is necessary to a radically relational perspective, because actual relations are only directly observable through concrete action. (Powell 2013, 194)

These quotes exemplify the broad trend toward conceptualizing social relationships and networks in terms of the processing of events, termed variously "interaction," "transaction," "practices," or "action." This variety is bewildering by itself. Often enough, the authors do not really discuss the implications of these concepts. For example, Mische refers to both "social practices" and "communicative interaction" in the same sentence, in spite of the fundamental differences of the two notions of practice and interaction. Often, these terms seem to be little more than shorthand expressions for the processual character of social life.

In addition, key authors of relational sociology disagree about the theoretical placement of these events and social ties in relation to each other:

- Jackson and Nexon envision a dual description of ties as processes *and* relations, akin to the wave/particle duality in physics (1999, 292). Emirbayer writes only about the process of transactions in the "Manifesto," without specifying how or whether these would harden to relations (1997, 28f). Dépelteau picks up on this, opposing the idea that transactions could stabilize in the form of relations or structures (2008, 2015). For him, to speak of "ties" is just a description of ephemeral events that do not build lasting structures.
- For White, social networks are first of all structures of meaning stabilized in the form of meanings and domains, and only changed by switchings bringing fresh sparks in (Godart and White 2010). Tilly (2002, 2005a) and Mische (2003) go further in arguing that interaction/transaction generates ties and networks, rather than merely changing them. Again, network ties and the processing of events seem to be two separate things, but with a stronger role for events.
- Azarian, Crossley, and Martin all write of social relationships and networks in terms of stabilized series of interaction. Ephemeral events lead to "expectations" (Azarian 2010, 326; Crossley 2011, 28) that guide future events. In a similar sense, Martin writes that a relationship is indicated by "the possibility of a specific type of interaction" and that "particular actions [are] appropriate" to a relationship (Martin 2009, 11).

We have thus three different approaches to the interrelation of events and networks ties:

(1) Events can be primary, networks and ties mere descriptions of them (Dépelteau).
(2) Events can be secondary, leading to the change of already existing networks (White).

(3) Events can generate and stabilize network ties, with these governing subsequent events (Azarian, Crossley, Martin).

Mische and Tilly seem to adopt a mixture of (2) and (3). Emirbayer and Jackson/Nexon are not quite clear about their positions. I personally advocate approach 3, with a strong role for social events and with social relationships and networks forming in communicative process.

In terms of philosophical foundations, this entails a turn away from the relational ontology laid out by Ernst Cassirer. Cassirer is frequently heralded as providing the foundations for relational sociology (Emirbayer 1997, 282f; Mohr 2011). He offers an ontology of the social world built on relations between substances, instead of focusing on these substances, as individualists or holists do. But these relations are themselves christened as foundational, as if the social world were composed of relations between substances. If this were the case, how can social relationships ever form or end? Where does change come from? Why is everything not held in place by the relations gluing it together? Instead, we can turn to the *processual ontology* of Alfred North Whitehead ([1929] 1978). The social world is first of all composed of events, not of substances or relations between them. In current theory discourse, Andrew Abbott (2016) prominently advocates processual ontology, but without discussing social relationships and networks.

Assumption 7.1. The social world consists of ephemeral communicative events.

As I will argue, not just any event forms part of the social world, only the *communicative* events where some forms of meaning (words, symbols) are processed with various actors (individual, collective, or corporate) involved. Such events can, but do not have to, crystallize into relatively stable structures, including social ties and networks. The problem of social order, central to social thinking since Hobbes, then becomes: How do ephemeral events stabilize into social structures and cultural patterns, including relationships and networks? This question reminds us of the structural functionalism of Talcott Parsons, who sees the maintenance of an ordered set of relations in the social world as imperative ([1951] 1964). In contrast, my concern is analytical. When, and how, do ephemeral events lead to relatively stable structures? And when and how do they change?

My answer picks up on Parsons, as well as on Weber and Luhmann, with the concept of "expectations" also advocated by Azarian and Crossley (as discussed earlier). Social structures are patterns of expectations that crystallize over sequences of events (see section 2.2). These events are observed and interpreted with regard to what to expect of future events. These interpretations and expectations solidify and guide subsequent events. Say Eric invites his colleague Sue to his birthday party. This contributes to an expectation of friendship between them, making it easier to switch from collegiality to amicable behavior in the future and harder to remain purely formal and collegial. Probably, Sue will now (have to) invite Eric in turn.

> Assumption 7.2. Social structures and cultural patterns form, stabilize, and change over the course of communicative events in the form of expectations. In turn, these expectations guide future events, making for their relative predictability and for the stability of expectations.

In a way, these two assumptions provide the most general foundations for the theoretical perspective in this book. Assumption 2.1 about the interplay of meaning structure and patterns of communication in social networks already builds on these two and details them for the case of social networks. Assumptions 7.1 and 7.2 start from a processual ontology of the social world as composed of ephemeral social events. And they argue that social structures and cultural patterns develop in the course of these events. The precise conceptualization of these events and their processing is advanced in the next chapter.

In a way, this interplay of events on the one hand, and social and cultural patterns on the other hand, constitutes a "duality" in the sense of Ronald Breiger (1974, 2009, 2010). Events and sociocultural patterns co-constitute each other. This duality is more fundamental than those of actors and groups, of structure and culture, or of cases and variables because it captures underlying social features and processes. From my perspective, structure and culture, actors and groups, and cases and variables are only co-constitutive because events and sociocultural patterns process in tandem. Also, the social structures and cultural patterns from assumption 7.2 do not only include network phenomena (as in assumption 2.1). Other kinds of sociocultural structures, including markets, states, institutions, formal organizations, and identities of social actors similarly form and develop over the course of communicative events, as well as guide these in turn (Figure 7.1).

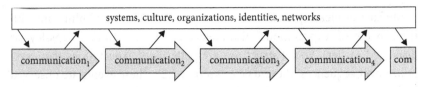

Figure 7.1 Co-constitution of communicative events and social structures/ cultural patterns.

Source: From Jan Fuhse, 2018, "Deconstructing and Reconstructing Social Networks," in: François Dépelteau, ed., *The Palgrave Handbook of Relational Sociology*, p. 463 (New York: Palgrave). Reproduced with permission of Palgrave Macmillan.

7.2. Behavior

How can we best conceptualize these events, and their stabilization in the form of social relationships and networks? The simplest way of looking at network processes is to conceptualize them as behavior. This perspective assumes that individuals behave wholly (or mostly) as triggered by external forces in a stimulus–response model. Aspects of meaning are bracketed in this approach, as laid out by George Caspar Homans (1961). It assumes that all individuals behave more or less in a similar way, given the external circumstances. These external circumstances (opportunities and restrictions) are given by the network: people act according to their position in a network without regard for cognition, individual preferences, or cultural imprints.

It is easy to see that this model conforms to the traditional approach in network research. Networks of social relationships are viewed as acultural structures determining individual behavior, at least to a large extent. The disregard for meaning and culture echoes that of much network research (see chapter 1). However, in this approach, the focus lies on the individual, not on the network. All social processes are reduced to individuals behaving in structurally conditioned ways, and these individual behaviors have to be summed up to lead to network structures and to social phenomena (springing from network structures).

This is akin to the approach of *social capital* that treats networks as resources for individuals (Lin 2001). The social capital approach disregards both the cultural forms interwoven with networks, and the different culturally given preferences of individuals (Burt 1992). Even James Coleman's work, often regarded as one of the hallmarks of action theory of networks, brackets meaning (1990a) or treats it as a mere correlate of network structure

(1990b). Rational choice is a close cousin to behavioral approaches, as long as it does not allow for individual variation of preferences (that does not correlate with structural conditions).

The concept of social capital, in combination with a rational choice perspective, takes three important steps (Fuhse 2020c, 37):

- It reduces social networks to the *opportunities or constraints* afforded to individual actors. Networks are seen from the perspective of the individual here, but only structurally as enabling or constraining their actions.
- The notion of capital implies that networks are a *resource* that actors can possess *more or less* of, like economic capital (money) or human capital (skills). Therefore, we should be able to quantify the value of relationships and networks for individual actors, or at least to compare their value.
- Finally, the social capital literature assumes that actors will try to *maximize the value* of their social networks (Lin 2001). New relationships are formed, and old relationships maintained or dropped, due to the costs and the benefits associated with them.

These steps make for a purely structuralist perspective on social networks. As in a simple stimulus–response model, individual actors are confronted with clear-cut situations, offering them rewards and punishments. These propel them to behave in a certain way, without taking different preferences or cultural imprints into account.

There is nothing inherently wrong with such a behaviorist approach to social networks. The concept leads to simple and straightforward models. Much fruitful research (see the works by Granovetter and Burt, among others) is implicitly viewing social processes as behavior, even if this is obviously counterfactual. People do behave differently in the same situation, given their varying cultural background. Accordingly, the recent "cultural turn" in sociological network research should lead to theoretical models with more explanatory power. Furthermore, while it is easy to see how networks lead to specific behavior, the networks themselves have to be assumed as givens in this perspective. Behavioral theory does not offer an account of what networks are, or how behavior leads to the formation, reproduction, and change of networks. Thus, it does not yield much surplus for the theory of social networks.

7.3. Action

Following the classic definition of Max Weber, "action" is a special instance of human behavior in that it is based on subjective meaning ([1922] 1978, 4ff). Merely reactive behavior (e.g., somebody instinctively ducking a flying object) would not count as action. This, of course, mirrors George Herbert Mead's discussion of instinct-led behavior as opposed to behavior based on the inner dialogue of an actor who asks herself about the meaning of objects for her and for others ([1934] 1967, 68ff). Mead insists that all meaning is the result of interaction with (mainly significant) others ([1934] 1967, 200ff). This offers a first link to networks: connections to friends and family lead to the subjective meaning on which actors base their decisions to act. This can be termed the "milieu effect" of social networks, since individual dispositions for action follow the imprint of their formative milieus.

This effect is included in Burt's framework for the interplay of social structures and action (1982, 9). Social structures (in particular: networks) do not only affect action directly, as opportunities and constraints, as with the concept of behavior (see earlier discussion); social networks also influence the interests that actors pursue, and these interests underlie the decisions for or against particular courses of action. An obvious instance of the milieu effect can be found in Karl-Dieter Opp and Christiane Gern's work on the role of networks for individual participation in social movements (1993). People influence each other in whether they hold particular grievances as important enough to engage in political mobilization.

The milieu effect is mainly based on the *composition* of an actor's personal network, and on the *cultural values and symbols* distributed in them. It leads to particular interests or desires that differ between actors with different cultural backgrounds or peer groups. The *structure* of a social network, in contrast, awards the individual actor with options for action due to the constraints and opportunities coming with her network position. This second effect of networks on individual action is discussed in the social capital literature and does not vary with subjective meaning (see earlier discussion). The challenge for an action-theoretical model of networks, now, lies in the simultaneous consideration of both of these effects—networks influence both an actor's opportunities for action and her evaluation of them (figure 7.2).

This renders ceteris paribus arguments tricky about what happens if somebody has more or less weak ties at her avail, or if somebody is brought up in

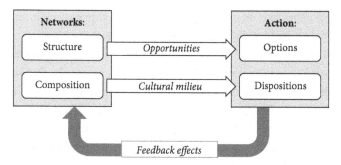

Figure 7.2 Action theoretical model of social networks.

upper class family or an ethnic minority group: both effects crosscut and lead to very particular outcomes depending on their combination. This can be seen in the research on migrant groups where dense network structures that often have positive effects on individual attainment sometimes show a "dark side of social capital" (Portes 1998, 15ff; see section 4.3).

Another challenge is to model the *feedback effects*: How does action (that is influenced by network structure and composition and by other social phenomena) lead to new network structures? Not all action is relevant, only the action termed "social" by Weber in that it is oriented toward the behavior of others ([1922] 1978, 22ff). In this vein, social networks are composed of the subjective orientations (sympathy, antipathy, esteem, etc.) toward significant others (that is, alters with a particular meaning for ego). These are the results of earlier interaction and lead actors to treat their alters in ways determined by the subjective meaning attached to them.

This is in line with Weber's definition of social relationships, as quoted in section 2.4 ([1922] 1978, 26f). According to Weber, social relationships consist of the meanings that a set of people ascribe to each other (and to their association) that lead them to act in a specific way. Where does this meaning of a relationship come from? From previous social action among the parties involved that has led to a set of expectations that characterize their relationship. Weber's definition actually encompasses all sorts of social constellations (including formal organizations). But it can be read as an action-theoretical account of network ties. Networks here consist of the set of subjectively held expectations about the behavior of others within a specified population.

With Johan Galtung, we can distinguish between normative and cognitive expectations (1959). Cognitive expectations denote what we think *will* happen ("predictions"), irrespective of whether we want it to happen.

Normative expectations instead capture that we think something *ought* to happen (or not happen; "prescriptions"). Social relationships combine these two types of expectations: We expect our friends cognitively to treat us in a certain way. But we also normatively want them to act thusly. To have a social relationship means not only that ego expects alter to treat her in ways typical for the relationship at hand (amicably, lovingly, etc.); it also means that ego *perceives* alter to hold similar expectations with regard to ego's actions. Properly speaking, then, individual action in a social relationship is based on *ego's expectations* (normative and cognitive) *of alter's actions*, but also on *ego's expectations of alter's expectations* (normative and cognitive). The relationship itself is a bundle of mutually attuned expectations and expectations of expectations from both alter and ego. The expectations on the two sides do not have to be symmetrical, and they can be in conflict as well as harmonious. But in the repeated back and forth in a relationship history, they will tend to complement each other, and to lead to an agreed-upon definition of the relationship. Action theory in principle has to model each of these steps in succession and examine how alter's actions affect ego's expectations, and vice versa.

However, true to the general theory of action theory, the key variable is not what ego expects to happen, but what she wants to happen—her interests and desires. We can model these as stable and assume that ego starts with an interest in a particular relationship with alter. This interest would then guide her actions irrespective of the actions and reactions from alter. This sounds unrealistic, and I have argued earlier that individual interests depend on her embeddedness in a network of relationships (the "milieu effect"). The existing relationship to alter, but also those to third parties, would affect ego's subjective desire for a particular relationship with alter. Strictly speaking, then, ego and alter influence each other in their interests in the relationship through the history of their repeated actions. But the more we treat this succession of acts as a flow of mutually oriented behavior, with an emerging definition of the situation that actors orient to, the more we conceptualize social events as "interaction" (see section 7.6). The model of alter and ego independently reasoning about their actions (on the basis of subjective interests and expectations) becomes too complicated to be useful.

Action theory thus has to offer a distinct and plausible account of what networks are, and why they stabilize and develop—even if I have not found it elaborated by action theorists, yet. In addition, it is able to translate networks into determinants of social action. Networks and relationships affect both the opportunities for action (as structures of expectations ego faces), and the interests of individual actors. However, there are three problems associated with this perspective:

- Action theory has to make assumptions about what people want or think—their cultural imprints and their motivations for action. These are "unobservables" that can be only be deduced post hoc from observable behavior (or from the situation and from actors' social environments). Weber's approach was to deal with this by attributing ideal types of motivation to actors on the basis of their observed behavior and on their involvement in social contexts like family, church, sects, bureaucracy, and economic life. Interestingly, not only sociological observers but also the participants in social processes themselves do this. They can only infer the underlying cognition and motivations leading actors to behave in a certain way. An action-theoretical account has to conceive of networks in terms of subjective expectations—but it also has to deal with the fact that neither we nor anybody else knows exactly what other people expect, want, and think. It might be better to start from the fundamental fact that individual thoughts and expectations are attributed and not known in social processes, as in chapter 8.
- As elaborated previously, action theory would have to model the intense back and forth in social relationships as a succession of independent actions. These are based on an acting ego's expectations and interests, but also on ego's expectations of alter's expectations, all of them potentially changing with every successive move (by alter and ego). This sounds complicated, and action theoretical models usually respond to this challenge by holding interests constant. They assume—counterfactually— that people know what kind of relationships they want with others *before* interacting with them.
- This already points to the pressing problem of aggregating the various actions in a network to new states of social structures. Let us assume we know the opportunities actors face and how they evaluate them. But then actors face *different* opportunities according to their network position (and diverge in their evaluations due to their individual social environments). Hence, they act differently, and these different actions have to be aggregated to a new social situation (Coleman 1990a, 20f). Opp and Gern's research only deals with the atomistic and individual decision to participate in a demonstration, not with the varying actions

of social movement organizers, peripheral members, bystanders, state authorities, etc. Peter Hedström advocates for agent-based modeling to solve this problem (2005, 67ff). However, this strategy up to now only covers a limited range of social phenomena. Action-theoretical accounts of processes in social networks are best suited to deal with undifferentiated structures, where individual variables like unemployment, attainment, or participation in a social movement are explained by network position and composition. It fares less well when it comes to modeling large-scale phenomena or the change of networks themselves.

The SIENA algorithm and software offer a sophisticated application of the action-theoretical approach to network data (Snijders, van de Bunt, and Steglich 2010). SIENA starts from data on relationships in full networks, for example, friendship choices in school classes, over a number of points in time (at least three). The model does not predict what kinds of network structures result from utility-oriented individual choices. Rather, it examines what kinds of individual preferences for particular relationships (homophily, reciprocity) and micro-constellations (transitivity, preferential attachment) would lead to the network patterns observed. These preferences are not modeled by individuals on the basis of observed choices but assumed to be constant in the population.

Individual relationships are interpreted as the result of two separate decisions for a relationship by the actors involved. If a two-sided friendship appears between alter and ego from t1 to t2, the model infers that one side decides that she wants to be friends with the other side, first, and then the other side reciprocates. The model also allows for one-sided relationships and even requires them as a necessary step to reciprocity. This seems odd when considering ties like friendship or love. In addition, the question looms whether two independent decisions make a relationship. In the approach favored here, relationship formation does not result from isolated preferences, but from a back and forth of communicative events between the actors involved. In any case, SIENA, in spite of the focus on individual decisions in its model, only deals with longitudinal series of cross-sectional snapshots of network data. While it asks what decisions could lead to these snapshots, it does not actually examine the process. However, SIENA takes the impact of individual choices (for or against relationships) on subsequent individual choices into account. This solves the third problem, in principle, even if it does so by inferring individual preferences from network structures, rather than the other way round.

Overall, the action-theoretical approach to networks suffers from its focus on the individual as bringing about relationships and network patterns. While this seems counterfactual with regard to singular relationships, the aggregation of choices into network patterns becomes a real problem. Also, the action-theoretical reconstruction of networks as patterns of expectations (and expectations of expectations) sketched earlier does not really enter the modeling of individual action in networks—be it in the social capital approach or with the SIENA algorithm. Without this connection, action theory remains an intuitively plausible idea that does not yet lead to a coherent account of what happens in networks.

7.4. Practice

Another candidate for modeling the basic social processes in networks is the notion of social practices. Practice theory was developed by a number of very different scholars, from Ludwig Wittgenstein through poststructuralists to Pierre Bourdieu (Turner 1994; Schatzki 1996; Reckwitz 2002). These authors suggest that people rarely think much about what they are doing, as the theory of action suggests. Instead of conducting elaborate calculations about how to best follow their preferences under given circumstances, people act out of habits—the *habitus* in Bourdieu's formulation. According to Bourdieu, habitus are:

> systems of durable, transposable dispositions [. . .] The habitus, a product of history, produces individual and collective practices—more history—in accordance with the schemes generated by history. ([1980] 1990, 53f)

Bourdieu views the habitus as the product of social fields (like education, art, politics, etc.) and in particular of the constellations in these fields. A bank clerk knows exactly what a bank clerk is supposed to do and acts accordingly, without thinking much about preferences or opportunities. Other authors do not really discuss the social constellations making for the habits leading to practices, or the social constellations resulting from them. Therefore, I focus on Bourdieu's theory of practices.

Bourdieu stresses the relational nature of social practices and of the underlying habitus. The practices of a peripheral artist acquire meaning only in relation to those of central artists, art dealers, art galleries and museums, critics, and all the other actors in the field (Becker 1982). These together "make" the world of art as a system of mutually attuned practices driven by an ordered plurality of individual habitus. Although these practices are relationally attuned, they should also be seen as continuously fighting each other for recognition, for symbolic prestige, and for the other resources that are distributed in the field.

Bourdieu emphasizes diverse resources (cultural, economic, and symbolic capital) determining the "objective" social relations that underly the individual habitus and resultant social practices. The positions in the field are defined by the distribution of these resources, and both habitus and social practices are prescribed by this "deep structure" (Bourdieu and Wacquant 1992, 71ff). Accordingly, the social practices and the relations among actors in a field would more or less follow the distribution of resources. The "relational" in Bourdieu's theory boils down to the "relative" construction of social positions according to their resources (Lash 1993, 201; Bottero 2009). He criticizes social network analysis as overly concerned with intersubjective relationships (rather than "objective relations" as the underlying structural causes; Bourdieu and Wacquant 1992, 88f). At another point, networks are reduced to individual resources in the form of social capital (1986). How this connects to his general theory of fields and social structure remains unclear.

In spite of this vague disregard for the importance of social networks, Bourdieu's theory of habitus and the other practice approaches offer important insights for the rooting of individual behavior in the social environment. In a nutshell, they argue that people seldom act out of rational calculation of means–ends relationships. More often, we simply follow the routines and the perspectives imposed on us by the social structures around us. When in Rome, we do as the Romans do. And when somewhere else, we rapidly switch to the respective behavior and frame of reference. The argument not only applies to social environments, but also to positions in these: a bank clerk "knows" how to treat customers, other clerks, his bosses (at various levels), and the cleaning personnel without having to think about it.

In this interpretation and bracketing the question of determination by resource distribution, Bourdieu's formulation illuminates how people fit into network structures, in particular into the role structures inherent in networks (DiMaggio 1992; Mohr 1994). In addition, it offers a model of how meaning is processed in social networks: Networks are composed of inter-personal expectations that make up a social field, and social practices serve as symbolic markers of these relations between actors. Thus, networks are enacted as relational structures of social practices. The individual correlates of these social processes are the habits/habitus, learned in network positions and adapted to them. More often than not, people follow the rules and expectations embodied in social contexts rather than continuously ponder them in rational calculation.

Accordingly, networks and fields are inert structures. They are reproduced by incorporated habits, driven by a sense of position on the part of the actors involved. This line of argument obviously works best for *individual* actors who are able to incorporate and to internalize practices into their bodily habits and their mind frames. Networks between *corporate* and *collective actors*, say, between political parties, administrative units, lobby groups, and social movement organizations, cannot really be traced to subjective dispositions and routines of these actors. Rather, they would have to form part of the habits and mind frames of the individuals involved, which then scale up to the behavior of the collectives and corporates (Emirbayer and Johnson 2008, 26ff).

In network research, we find numerous examples of habitual action. Christopher Ansell's study of the late 19th-century French labor movement shows that the common practices of the "bourse du travail" was more important in uniting the movement than ideology (1997). Peter Bearman argues that the network of generalized exchange of women among segments of an Aborigine tribe is only upheld by the actual exchange practices (and not known by the participants; 1997). The closest application of Bourdieu's theory can be found in Helmut Anheier, Jürgen Gerhards, and Frank Romo's study of networks among writers in Cologne, Germany, and the distribution of various forms of capital (social, cultural, economic, symbolic; 1995). They demonstrate that the role structures (of blockmodels) in the writers' networks correspond to the distribution of these various types of resources. However, their theoretical interpretation integrates social networks much better than Bourdieu's own theory.

Linking social networks to a theory of practices yields considerable convergence and advantage. However, such a combination would be mainly geared toward accounting for the conformity of individual practices to established social structures. This explains the stability of networks, not their change. Networks remain static; their rearrangement would have to be traced to underlying changes in deep social structures (such as the distribution of resources) or to individual errors in following the structurally prescribed practices. Another question is: If people simply follow suit, why should we actually focus on individuals as the unit of analysis? Would it not be more parsimonious to model the social forces in fields and networks without recourse to individuals?

Many authors locate the driving force for practices in mind frames such as the habitus (Bourdieu) or "habits" (Turner 1994). Conversely, Theodore Schatzki asserts that the human mind and action are constituted by social practices (1996). Practices and subjective meaning co-constitute each other, but we really know about subjective meaning (habits or habitus) by inferring it from practices. How can this philosophical reflection be analytically useful? Bourdieu hypothesizes the habitus to follow the position in social structure. In Bourdieu's account, the social structures (and, in particular, the distribution of various types of resources) are doing the main explanatory work, and the actors are by far not as agentic as many sympathetic observers would have it. The individual might act as the unit of analysis (e.g., in correspondence analysis). But she is not really seen as "causing" social patterns, including networks.

7.5. Exchange

The first three concepts of social processes discussed so far focus on individual behavior and aim at explaining either individual behavior from network constellations or the other way round (or, ideally, both). The remaining five concepts—exchange, interaction, transaction, switchings, and communication—start from the interpersonal. The exchange perspective originally developed out of behaviorism, with the works of Homans a core reference of both approaches (Emerson 1976, 335). The perspective was fleshed out by Peter Blau and connected to relations and networks by Richard Emerson and Karen Cook.

Blau starts from the assumption that human beings depend on gratifications of their actions from others (1964). Consequently, people develop expectations about whom to get valuable gratifications from. But every reward bestowed also creates obligations to reciprocate. People giving out many gratifications are bestowed with power, because many people depend on them (with obligations to reciprocate), and granted status (as one way of reciprocating rewards). The back and forth of mutually gratifying actions and obligations leads to social relationships, such as love (Blau 1964, 76ff; see section 6.9). Social relationships thus feature both cognitive expectations (who will I get gratifications from?) and normative expectations (who is obliged to give me gratifications?), in the sense of Johan Galtung (see section 7.3).

Emerson uses exchange theory to explore the dynamics of network constellations (1962, 1972). Exchange here builds on existing network constellations, but it also changes and stabilizes them. Actors with favorable network positions are able to impose exchange to their advantage on others. At the same time, preferred exchange partnerships tend to emerge over time, as exchange leads to stable expectations between actors (Cook and Emerson 1978, 734f, 737). Later elaborations address the dynamics and stabilization of network constellations out of the desire of actors to attain favorable exchange ties (Willer et al. 2002, 116ff) and out of the perceptions and affections connected to network positions (Lawler, Thye, and Yoon 2008), and propose combining network exchange with processes of identity formation and negotiation (Burke 1997).

Exchange theory produces an innovative take on network constellations, and I am unable to review and discuss the wide variety of conceptual and empirical work in this tradition here. Overall, exchange theory remains rather abstract, with a desire to formalize social events, relationships, and network constellations. There is much to like about this strategy—abstraction and formalization contribute to the success of exchange theory. However, the approach reduces events, relationships, and networks considerably, to patterns of valued ties. Much interaction cannot immediately be mapped as the giving and taking of gratifications, as exchange theorists themselves concede. But is quasi-economic exchange really more important to relationship formation than chitchat? Or can we formalize a wide variety of events as exchange with valuations? Exchange theory relies on idealized models of social constellations. And it does not really deal with actual social process, apart from artificially reduced interactions in experimental settings (e.g., Cook and Emerson 1978).

Consequently, network exchange theory does not distinguish between different kinds of interaction and relationships. Friendships, love, and family relations, even business partnerships, are similarly treated as ties with a great deal of exchange and valued goods. We learn little about what distinguishes these kinds of relationships, or how we deal with them in actual social events. The "meaning" of relationships is dealt with cursorily, as obligations (normative expectations) resulting from previous valued exchanges. Where are these obligations located? Exchange theory determines them objectively by looking at structural constellations. They are not held subjectively by individuals, but also not negotiated between them. Much like behaviorism and rational choice theory, exchange theory offers an *as-if model* of the social world. More than the concepts of behavior, action, and practices, it focuses on what happens between actors, rather than in their minds. But it does so bracketing meaning and most of the actual processes at play. Therefore, it is unable to serve as a foundation for the examination of the construction of networks in social events, and for the arguments about the interplay of network constellations and forms of meaning from chapters 2 to 6.

7.6. Interaction

The concept of interaction comes from the tradition of the Chicago School of William Thomas, George Herbert Mead, Herbert Blumer, and Erving Goffman. A number of relational sociologists pick up on the concept (as does Abbott; 2016). Following the classic formulation by Mead, interaction can be conceptualized as an interpersonal process between at least two actors that is inherently based on the internal processing of meaning (symbols) by the participants ([1934] 1967; Blumer 1969). In this vein, the internal dialogue, including asking oneself about the meaning of symbols for others and taking the role of the other, is part of the interaction process. This strong focus on the processing of meaning distinguishes interaction from exchange, as discussed earlier.

What is the difference between this social psychological concept of interaction and Weber's concept of social action? First, symbolic interactionism does not follow Weber's method of classifying individual motivation into ideal types but insists on examining (or inferring) actual cognition. Second, symbolic interactionism stresses the interactive nature of social processes: the focus does not lie on the individual, but the individual and her mental processes are seen as part of a larger process of interaction involving at least two individuals (and the symbolic exchange between them). I have argued previously that social relationships are hard to conceptualize as a series of isolated individual actions, with the underlying preferences changing based on previous actions and on the definitions of the situation offered (see section 7.3). This back and forth between alter and ego and the developing patterns of subjective and intersubjective meaning (preferences and definitions of the situation) are better captured as symbolic interaction, in the line of Thomas, Mead, Blumer, and Goffman.

If we follow this lead, networks consist of the supra-personal processes (of exchange or communication) in the ties as well as of the internal (mental) processes in the nodes. Of course, this complicates things in comparison to models that focus exclusively either on internal processes (action) or on distinctly social processes (communication or transaction, to be discussed further). Here we face a tradeoff between the parsimony and the comprehensiveness of our models. The integrated perspective of interaction is less parsimonious, but it offers something for everybody: for those who insist on the autonomy of individuals, and for those who believe in the supra-personal dynamics of the social process.

Symbolic interactionists stress these two aspects of symbolic interaction—the subjective and the intersubjective—differently. Authors like Mead and Blumer adhere to a social psychological perspective and pay close attention to subjective processes (including their relation to social processes). This line of reasoning develops similar arguments to action theory (see section 7.3). Other authors, like Erving Goffman ([1959] 1990, 1981) or George McCall and J.L. Simmons ([1966] 1978), focus more on the supra-personal dynamics of communication. Their claim is that people often hold up and follow a "social definition of the situation" even if they disagree with it. In this view, social processes are mainly determined by prior social processes, and less by mental processes. McCall and Simmons, in particular, emphasize that individual identities—what people are seen as in their social environment—are the result of these supra-personal processes. This second line resembles the communication theoretical perspective (see section 8.2 for a more detailed discussion).

There are a number of attempts to wed a symbolic interactionist perspective with sociological network research:

(1) Gary Alan Fine and Sherryl Kleinman argue for this combination in an early and important contribution (1983; see also Stebbins 1969). According to them, social relationships (as the basic building blocks of networks) are based on the meaning they have for the people involved. Since this meaning is always in flux, networks themselves are dynamic. With this basic conception, network analysis would be able to overcome its structuralism, and to conceive of networks as

dynamic and not static. For symbolic interactionism, the combination with network analysis offers a way out of its fixation on social groups (that are presumed to be homogeneous and bounded) and to turn to interrelational social structures that cannot be described as groups (see section 3.1). In spite of its intuitive appeal, the paper by Fine and Kleinman has received little attention and has not led to a reorientation of either of the two approaches.

(2) A very basic account of interaction in networks is adopted in Kathleen Carley's constructural theory (1986, 1991). Carley models interaction as the exchange of information, with information as a broad category for cultural forms (including values, etc.). People in contact with each other have a high probability of sharing a piece of information that only one of the exchange partners possesses. As a consequence, the stock of information (that is, the subjective meaning) of one of the interactants is changed. This, in turn, leads to a higher probability of interacting with people with similar stocks of information (homophily) and therefore to a changed configuration of the interaction network. Initial distributions of cultural forms make for the change of network structures and their alignment along cultural differences (Mark 2003). This approach resembles the work on exchange networks discussed earlier. Even if constructural theory deals with forms of meaning, both the exchange of information and the mental processes of the actors involved are modeled as simply and parsimoniously as possible. Constructural theory has obvious merits, some of them precisely because of its simplicity. The challenge will be to come up with a more accurate account of social processes in networks leading to the study of these processes.

(3) Randall Collins's work can be interpreted as applying a Goffmanian perspective to social networks. His research on networks in the history of philosophy (1998) does not turn toward the micro-modeling of social events in networks. But in *Interaction Ritual Chains* (2004) he offers an impressive account of social micro-processes that combines Goffman's ideas about interaction with those of Durkheim on the emergence of symbols in rituals. In repeated face-to-face encounters, people become emotionally entrained (either with positive emotional energy or in conflict with each other). Out of the mutual focus, symbols emerge that are laden with energy and acquire their meaning in the

context of the social structures out of which they arise. Consequently, networks are woven around symbols and produce emotional energy for their participants (and stabilize, as long as they do).

In his classic *Conflict Sociology*, Collins devises social structure as consisting of the "interactional side" of "real connections among people" and of the "ideational side" of social symbols and meanings ([1975] 2009, 71). He considers various genres of communication (advice giving, gossip, intellectual discussion, entertainment, etc.) as making for the connections between people but only hints at the differences between kinds of relationships ([1975] 2009, 74). Neither Collins nor Carley's constructural theory address the meaning of social relationships, the negotiation of these "definitions of the situation" in interaction, or its effects on network structure. While their approaches could be extended in this direction, they deal with ties mainly as channels of interaction, not as themselves imbued with meaning.

(4) A recent attempt to combine symbolic interactionism and network research comes from Nick Crossley (2011; see section 1.4.d). Crossley builds on Fine and Kleinman's work and similarly argues that networks are based on the subjective meaning associated with others, arising in interaction with them. Crossley's basic model features two individuals who arrive at a common ground and shared rhythm (following Merleau-Ponty) in "dialogue" (Crossley 2011, 28ff). Social relationships and networks are therefore mostly about breeding community and commonality in perspectives and cultural forms. This basic model accounts for the emergence of group cultures in dense network clusters, like local punk scenes and punk styles (Crossley 2015a; see section 3.5). Conflict and asymmetric role relationships (and structural equivalence) are not really covered.

While Crossley suggests incorporating qualitative methods (focusing on meaning) in network analysis (2010), Wouter de Nooy claims that it is possible to study symbolic interactionist conjectures about meaning with formal network analysis (2009). Andrea Salvini similarly proposes a combination of the structural perspective from network analysis with the interpretive stance of symbolic interactionism (2010). But he also argues that the approaches offer distinct perspectives that should not be conflated, or one subsumed under the other.

Overall, these various contributions point to a possible fruitful combination of symbolic interactionism with sociological network research.
However, the micro-modeling of how exactly interaction leads to the buildup
of networks, what the relation between mental and social processes is, and
how we should study them remain vague. In particular, the concept of interaction includes assumptions about mental processes that remain unobservable in sociological research. We can only infer these mental processes
but not tackle them directly. Also, the simultaneous attention to mental
and communicative processes, while theoretically convincing, is quite demanding methodologically.

7.7. Transaction

Mustafa Emirbayer introduces the notion of transaction (or "trans-action")
prominently into theories of social networks in his "Manifesto for a Relational
Sociology" (1997). This is taken up by Charles Tilly and François Dépelteau
but comes with very different theoretical implications for the three authors.

Building on John Dewey and Arthur Bentley, Emirbayer contrasts relational approaches that deploy a transaction perspective on the social world
with other approaches starting from "self-action" or "inter-action."

- Methodological individualism, including rational-choice and micro-
  sociology, but also systems theory start from a *self-action* perspective
  (Emirbayer 1997, 283ff). They view closed units (individuals, systems,
  societies) as driving social processes out of internal impulse.
- The *inter-action* approach, instead, focuses on the action taking place
  between entities. But importantly, the entities themselves remain intact,
  unchanged, through the process (Emirbayer 1997, 285f). Emirbayer
  sees the variable-centered approach as exemplifying this approach.
  Here, "fixed entities with variable attributes" are brought into statistical
  models to create outcomes.[1]

---

[1] Symbolic interactionism does not match this rendering of the "inter-action approach." The individuals participating in symbolic interaction surely change in their subjective meaning as the result of interaction. Conversely, we have to ask whether statistical models in the variable-centered
approach really feature any kind of action between entities. Here, variables are seen as causing (or
at least correlating with) other variables across individual cases. But the cases themselves remain
isolated, closer to a self-action approach than to any kind of inter-action between them. Perhaps, exchange theory could be termed an "inter-action" approach in the sense of Emirbayer.

- The "*trans-action*" approach heralded by Emirbayer sees the processes between entities as creating the relevant outcomes, including the change of the entities themselves. He writes:

> In this point of view, which I shall also label "relational," the very terms or units involved in a transaction derive their meaning, significance, and identity from the (changing) functional roles they play within that transaction. The latter, seen as a dynamic, unfolding process, becomes the primary unit of analysis rather than the constituent elements themselves. (1997, 287)

Emirbayer christens a number of theorists as "relational sociologists" who follow this transaction perspective, including Marx, Simmel, Cooley, Elias, Goffman, Foucault, Bourdieu, Collins, White, and Padgett, partly even Durkheim and Parsons. However, the notion of trans-action remains rather vague in Emirbayer's "Manifesto." And he does not detail its connection to relations or relationships. "Relational sociology" is here about process, "relating," rather than about "relations" or "relationships." This conceptual decision is not taken up in Emirbayer's other writings, where he turns to individualist concepts like "agency" and "practices," but also to field theory (Emirbayer and Goodin 1994; Emirbayer and Desmond 2015). Unfortunately, the theoretical option of transactions is not fleshed out here. But Emirbayer's "Manifesto" is clearer than works of most authors he calls relational sociologists (including Elias, Bourdieu, and White) about the fundamentally processual nature of the social world.

The original formulation of the concept of transactions by Dewey and Bentley ([1949] 1973) does not help much in this regard.[2] Their essay *Knowing and the Known* does not deal with social process as such, but with the fundamental problem of how to conceptualize the relation between scientific observers and what they observe. The term "transaction" here stresses that both the scientists and the objects of inquiry are affected by the process of observation. Accordingly, they oppose sharp distinctions between actors and their doings, arguing that both can be observed only in tandem:

> *Transaction* is the procedure which observes men talking and writing, with their word-behaviors and other representational activities connected

---

[2] For different interpretations, see the contributions in Morgner (2020).

with their thing-perceivings and manipulations, and which permits a full treatment, descriptive and functional, of the whole process, inclusive of all its "contents," whether called "inners" or "outers," in whatever way the advancing techniques of inquiry require. (Dewey and Bentley [1949] 1973, 137, original emphasis)

In some regards, this can be read as a precursor to science and technology studies and actor–network theory (ANT), with their close observation of scientists and scientific objects influencing each other. But what exactly do we learn about the process of "transaction"? As a first rule, Dewey and Bentley formulate:

consider everyday behavior [. . .] without subjection to either private mentalities or particulate mechanisms. Consider closely and carefully the reports we make upon them when we get rid of the conversational and other conventional by-passes and short-cuts of expression. ([1949] 1973, 141).

That Dewey as a pragmatist author calls for the disregard of subjective meaning ("private mentalities") is surprising. But also "particulate mechanisms" are to be avoided in our descriptions. Dewey and Bentley want to focus on observable processes, rather than assume too much (in "conversational by-passes and short cuts of expression") about their regularities or the subjective dispositions underlying them.

Dewey and Bentley then illustrate their point with a billiard game ([1949] 1973, 142). The game itself can be "profitably presented and studied interactionally," by considering how the movements of the balls affect each other. But "a cultural account of the game" would have to reflect the understandings connected to it, including conventions like "winning" and "playing for money." The transactional mode of observation here takes the "naming" of things into account, the meanings attached to observable behavior. However, it seems that these meanings are not considered subjectively or even in their intersubjective negotiation (as action theory and symbolic interactionism would do). The meanings seem to be inscribed in the doings and not reflected upon, as in the theory of practices. However, transactions are not traced to individual dispositions in the sense of habits or habitus. Dewey and Bentley's concept of transactions would be a practice theory without attribution to single actors. Their transactions are distributed across entities, much as in ANT (Latour 2005).

This original impetus of the transaction concept is taken up by Dépelteau in his "deep relational sociology" (see section 1.4.d). He argues against a "detached 'Subject' observing an 'objective' reality" (2015, 48). Actors and the processes of transactions between them are entangled and unseparable, and therefore Dépelteau rejects any dualisms of actors and their actions, of agency and structure, but also of transactions and the structures emerging in them and governing them. Social structures "that we call market, wedding, war, genocide, racism, exploitation, domination, love [. . .] should be studied as chains of trans-action" (Dépelteau 2008, 62). This is radical processualism, not in the sense of tracing everything to processes, but in the sense of reducing everything to process and denying that there are lasting features of the social world, including social relationships and networks. How exactly "chains of trans-action" like markets, racism, exploitation, domination, or love stabilize, or reproduce to form recognizable patterns, remains obscure.

This basically forbids any reflection on social patterns (by conceptual fiat!). To me, it looks like throwing the baby out with the bathwater. Here, the "relational" approach denies any forms of relationships, and it boils down to rules for observing social processes without theoretical abstraction, as ANT asks us to. Indeed, Dépelteau similarly rejects a distinction between human actors and material objects in our observations:

> we should avoid any unnecessary analytic distinctions. We should stick to the only "level of reality" we can see and know, which is made by complex transactions (or associations) between various human and non-human transactors. (2015, 49)

If the notion of transaction covers all sorts of influence between human beings and material objects, including the physical impact of one object on another, then not much more can be said about transactions. And just as the bat and the ball do not form a lasting relationship, human beings are forbidden from having them. Dépelteau's deep relational sociology not only gives no answers to the questions "How do relationships stabilize and change? And how can we study them?"; it also demands that we should not ask the questions.

If Emirbayer leaves open the nature of social relationships, and Dépelteau denies their existence categorically, Tilly takes a third option. He seems to have adopted the notion of transactions from Emirbayer, his junior colleague at the New School of Social Research in the 1990s. As cited in section 7.1,

Tilly sees "interpersonal transactions [. . .] accumulate into durable social ties" (2005a, 7). Identities, social boundaries, but also social relationships are the result of transactions, not only their descriptions (as Dépelteau would argue). First, on the level of observations, we infer a relationship from observing "a series of transactions between sites [. . .]: a friendship, a rivalry, an alliance, or something else." Second, and underlying these regularities of transactions, Tilly postulates:

> individual and collective dispositions result from interpersonal transactions. [. . .] Cumulatively, such transactions create memories, shared understandings, recognizable routines, and alterations in the sites themselves. (2005a, 7)

These few and rather sketchy formulations again point to a connection between inner (for individuals, mental) processes in nodes and the transaction processes in the relation. The concept of transaction, however, brackets the inner processes as secondary to the observable processes between nodes (or sites, in Tilly's formulation)—even if the regularity of transactions in relations (friendship, rivalry, etc.) derives from the creation of "memories, shared understandings, recognizable routines" within the sites. In contrast to the concept of interaction, the very processes coupling the interpersonal with the inner-personal are only vaguely hinted at in this perspective. Also, we know little more about how exactly transactions concatenate into identities, relations, and social boundaries (as Tilly's basic building blocks of larger social structures).[3]

Overall, the concept of transactions remains ill-defined and not clear enough to ground a theory of social networks. Dépelteau sees relationships as chains of transactions, without specifying how they stabilize (and rejecting the idea of a structural level over and above observable transactions). Emirbayer does not really relate the process of transactions with networks. Tilly comes close to my position when formulating that transactions lead to social ties (and boundaries and identities). Probably, he saw transactions as the driving process, and individual minds as carrying the meaning attached to social relationships. Even if we disregard this diversity in theoretical orientations, the link between transactions and meaning is not fleshed

[3] In late 2007, probably half a year before his death, Tilly remarked that he still wanted to write a book about how all social structures spring from transactions (personal communication).

out. Dewey and Bentley envision transactional accounts to incorporate the level of culture, but not in the sense of individual understandings and dispositions. Tilly seems to argue otherwise. Perhaps we can conceptualize the transactions themselves to carry meaning, and to attach it to social entities (actors) and structures (boundaries and networks). This would lead to a communication theoretical perspective (see further discussion). But it would also require much more theoretical detail.

7.8. Switchings

A final candidate for the social events in networks is the concept of "switchings," as introduced by Harrison White in several papers from 1995 onward (1995a; Mische and White 1998; White et al. 2007; Godart and White 2010). The concept is both structural and cultural. It denotes a communicative shift from one network to another (1995a, 1038), as well as a change in cultural forms (speech register; Mische and White 1998, 701). Invoking a new set of cultural forms in communication constitutes a switch in network ties, since these are intricately interwoven with, and defined by, cultural forms. For example, two colleagues may discuss an important work task but then start talking about private issues (or about sports; Erickson 1996). The formal network of work relationships (including its status hierarchies) is suppressed here, making way for the informal network of friendship relations in the company.

According to Frédéric Godart and White, such "switchings" make for the emergence of new forms of meaning (2010). The communicative shift activates different sets of social relationships interwoven with cultural forms, and this brings "fresh meanings" into each set. Generally, White seems to devise network domains as relatively stable sociocultural formations that tend to reproduce themselves. But the process of switching brings irritations, new combinations in. In our example from the previous paragraph, we can think of an intern impressing employees and superiors with his wit and profound knowledge of sports. This will make it easier for him to make inroads into the inner circle of the company, increasing his chances of being hired and promoted. At the same time, the collegial discourse about sports is shaken up and infused with new ways of thinking. Both the networks and the domain change.

It is easy to consider these kinds of sociocultural switchings as special cases of interaction that otherwise proceeds within the established lines of network domains. But what exactly constitutes a "switching," and what is not covered by the concept? In our example, the intern joining the established lunch network (without a change in topic, or cultural domain) makes for a switch, because the social network is changed. The collegial chat changing from work issues to sports—without a change in personnel—is also "switching" because a new domain is invoked. In an interview, White rejects the idea that switchings could be a special instance of communication. Rather, switchings are the general process of social events, and communication a special case of switchings (Schmitt and Fuhse 2015, 182).

Overall, White's concept of switchings offers a fresh perspective on communication. Switchings combine the structural side of network relationships with the cultural side of forms of meaning. Individual dispositions and cognition are bracketed here, focusing on the communicative practices. And White insists that we should not observe isolated social contexts but look at how they influence each other in micro-processes. The precise mechanisms at play remain underdeveloped, though. In particular, White focuses here on the change of existing formations. But he does not really elucidate how network formations come about, or how they reproduce.

7.9. Desiderata

What would an ideal model of micro-processes in social networks look like? The discussion so far has identified a number of core problems in conceptualizing the basic events that lead to the formation, stabilization, and change of social networks. Some of the requirements are as follows.

First, ideally, it should be possible to model theoretically how micro-events congeal to relatively stable networks (that can nevertheless change slightly or even dramatically with every new micro-event), but also how networks make specific micro-events more likely than others. Some of the concepts discussed are able to deal with the second causal influence (from network to micro-events), but not with the first. The behavioral perspective does not account for the formation or change of networks or even for what social networks *are* on the level of sociological theory. The related social capital approach reduces networks to individual resources, much as economic and human capital. Social practices will mainly lead to stability, not to the change of networks.

The concepts of action, interaction, and exchange fare best with regard to this point. Symbolic interaction offers and negotiates definitions of the situation, including definitions of relationships. These, in turn, provide starting points for subsequent interaction. Action theory could conceptualize networks as expectations, arising out of previous actions and affecting future actions. In practice, the back and forth of action with the continuous change of opportunities and interests is hard to model as a series of independent decisions. Exchange theory has a clear notion of social relations as obligations. Emirbayer, Tilly, and Dépelteau do not quite agree about how the process of transactions leads, or not, to network ties as relatively independent social patterns. White's switchings only account for changes of networks as sociocultural formations, not for their emergence or stabilization.

Second, in line with the perspective of relational sociology, social networks are inherently interwoven with meaning (symbols, schemes, expectations, categories, etc.; see chapter 2). The theoretical model of micro-events should be able to incorporate this. The concept of behavior abstracts from culture and meaning and does not account for their interplay with networks, as does exchange, mostly, apart from the very basic assertion that exchange creates obligations. Again, action and interaction meet this requirement. Both of them refer to subjective meaning as a core aspect of networks (e.g., the meaning of a relationship for the actors involved). Interaction also points to the processing of meaning on the social level, especially in Goffman's and in McCall and Simmons' conceptions. Transactions similarly seem to be infused with meaning, but probably more on the social side than subjectively. White's switchings combine the social-structural and the cultural, without recourse to subjective meaning.

Third, how should we deal with mental dispositions and processes? Both action and interaction refer to mental processes as the core mechanism linking social situations to micro-events. Social networks would lead to particular micro-events because of the subjective meaning actors hold of the relationships between them. We neither know much about these mental processes nor are able to observe them in social research. Weber's solution was to infer ideal-typical motivations to individual action on a post-hoc basis. While this may be convincing in many cases, it also points to the difficulties of including unobservables in theoretical models. Indeed, one of the problems of processes in social networks is precisely that not even the participants themselves are able to look into each other's head. If possible, a theoretical account of these processes should deal with this basic problem. This is partly done in interactionist theory (with processes like role taking or internal dialogue), but then it assumes that we as observers can better determine mental processes than interaction partners.

Fourth, the most promising avenue to avoid theorizing unobservables and to arrive at a model of micro-events that fits well with sociological network research is to focus on observable processes *between* actors. This option is pursued in approaches centering on exchange and partly in interactionist models. The concepts of transactions and of switchings aim in a similar direction but remain vague. In contrast to exchange, they incorporate meaning but without assumptions about mental processes, as in the concept of interaction.

Fifth, an ideal conception of social micro-processes in networks would provide clues for studying the formation, stabilization, and change of social networks. A sound theoretical account helps with the interpretation of existing studies. But one of the strengths of relational sociology is the strong link between theoretical reflection and empirical research. A theoretical account of micro-processes in networks should therefore also devise theoretical expectations for empirical studies and point to suitable methods.

We can roughly sketch the seven candidate concepts for social events (with communication as the eighth concept) as in figure 7.3. The vertical dimension covers whether the concept starts from the individual or regards social events as supra-personal. And the horizontal dimension reflects the importance that these concepts accord to meaning. Behavior and action

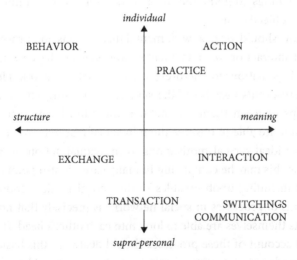

Figure 7.3 Relative positions of concepts for social events.

are individualist concepts that reject the idea of supra-personal properties of events. While *behavior* is individual and mainly structural, *action* theory incorporates a concern for subjective meaning. But mostly, action theorists decide for a reductionist modeling of meaning (e.g., in ideal types), placing this perspective a bit further to the structural pole than interaction, switchings, and communication. The notion of *practices* has a limited concern for meaning (as mostly derived from structural positioning), and it is less individualist than behavior and action, since practices and habitus are often tied to social constellations, rather than strictly to the individual.

The two notions of exchange and interaction are located between the individual and the supra-personal. Both place more emphasis on the processes between individuals than the concepts in the upper half. But the concept of *interaction* asks us to consider the (actual) processing of meaning both in individuals and in the interaction between them. And the *exchange* perspective ties its models of supra-personal processes to individual calculations, almost as in a behaviorist perspective. Meaning is incorporated only cursorily here, as obligations for exchange (normative expectations).

Transactions and switchings are located near the supra-personal pole of the diagram, as is the concept of communication (that I elaborate upon in the next chapter). All three of them relegate individuals to secondary importance, focusing on the processes between them. The *transactionists* remain vague about the importance of meaning, and how we could examine it. White, in contrast, stresses the social and cultural side of *switchings*, even if the concept is underspecified in other regards. Like switchings, Luhmann's theory of *communication* is firmly placed at the supra-personal and the meaning poles of the diagram.

7.10. Conclusion

These considerations lead to an unusual approach to social micro-events: Instead of focusing on individual actors and their subjective understanding of a situation, their motivations, and their behavior, micro-events could be conceptually detached from the actors involved. In this strategy, we would locate meaning in the micro-events themselves. It is fairly easy to see how an utterance can carry meaning, irrespective of the intentions of its sender or the understanding of the recipient(s). In this model, the meaning of one utterance would connect to the meaning of previous and of subsequent utterances. Social networks would have to be regarded as structures of meaning arising out of the flow of transactions. And as Emirbayer and Tilly formulate, the identities of the actors in a relation would be negotiated and established in relation to each other in this sequence of micro-events (as part of the meaning structure of social networks).

I argue in the following chapter that Luhmann's theory of communication ([1984] 1995, 2002)—with a few modifications—provides almost perfect foundations for the processing of social events in networks. According to Luhmann, all social phenomena develop in the course of communication. As in the theories of action and of exchange, social structures (including networks) consist of expectations. These derive from previous events and structure subsequent events. But these expectations are not (solely) to be found on the subjective-cognitive level of individuals. Rather, they can be conceptualized as provisional definitions of the situation (as in symbolic interactionism) established and continuously renegotiated in communication. Social events, then, carry meaning themselves, rather than merely signify something to the individuals involved.

Tom says to Betty, "I love you." This event has a certain meaning, in spite of neither Betty nor academic observers knowing exactly what Tom subjectively meant with it (see section 6.2.c). Similarly, we cannot determine how Betty understands the utterance, and neither can Tom. Participants in communication usually examine the meaning of the communicative event itself and then look for signs of subjective intentions and understandings in the communicative events following up on this sentence, and in those leading to it. The succession of communicative events (with the meanings offered by them) becomes the primary focus. Social relationships and networks build relatively lasting structures out of the general uncertainty inherent in all communication (see section 2.4). We relate to each other, build friendships, love relationships, and love relations precisely because we cannot look into each other's heads—because we want to know what to expect of others. Some of this may sound unusual and counterintuitive. I hope the readers bear with me to see where this approach leads.

8

Networks from Communication

In the last chapter, I argued that an ideal conceptualization of network processes should meet the following demands:

- It accounts for the effects of networks on events, but also for the effects of these events on network change.
- Following the perspective of relational sociology, networks are seen as meaningful social constructs.
- We should avoid assumptions about unobservable mental dispositions and processes as the driving factors of processes in networks. Neither academic observers nor participants are able to look into each other's heads.
- The change of networks is conceptualized as realized in the processes *between* actors, rather than by individual behavior.

This chapter argues that the concept of communication leads to a consistent account of the basic processes in networks in line with these requirements.[1] I take the concept from Niklas Luhmann but add important changes. Social relationships and networks arise out of communication and effectively structure future communication. This takes us to the heart of the framework from section 2.1: Relational expectations as the "meaning structure of social networks" arise, solidify, and develop over the course of communication. And they guide communicative events and make for a certain predictability that we observe in the "patterns of communication" between actors in a network. While these arguments could be made with other concepts like social action, interaction, or exchange, the concept of communication illuminates this interplay in a theoretically and methodologically fruitful way. I argue in the following that a modified version of Luhmann's theory of communication gives answers to the following questions:

---

[1] The chapter has been revised and extended from a previous version published in the *European Journal of Social Theory* 18 (2005). I thank Sage Publishing for the permission to use the article again.

(1) What exactly are social networks, and where do they come from?
(2) How do networks change, and why should they be relatively stable?
(3) Why should social networks be central mediators of social processes, as a lot of research testifies?
(4) Can we study and interpret networks between collective and corporate actors (like firms, states, or street gangs) in the same way as networks between individuals?

Some relational sociologist authors already pick up on Luhmann's concept of communication. In a recent article, White claims Luhmann's theory of meaning and communication to be compatible with his theoretical account of networks (White et al. 2007). In his studies on turn-taking in manager meetings, David Gibson cites Luhmann's concept of communication approvingly (2003, 1357). Luhmann plays a much more central role in Stephan Fuchs's theory of cultural networks (2001a). But even White's "switchings" (Godart and White 2010) and the concept of "transactions" used by Mustafa Emirbayer (1997) and by Charles Tilly (2002, 2005a) show important similarities with Luhmann's theory of communication: the sequence of social events is seen as a "dynamic, unfolding process" (Emirbayer 1997, 287), and little importance is accorded to mental states and processes (see sections 7.7 and 7.8). White, Tilly, Gibson, Fuchs, and Emirbayer (in his "Manifesto for a Relational Sociology") focus on the processing of meaning in the social events taking place between actors. Yet, none of these authors offer a detailed and consistent account of these processes, as I propose in the following pages.

I first discuss early formulations of the concept of communication (section 8.1) and the similarity with Goffman's concept of interaction, with conversation analysis, and with Foucault's notion of discourse (section 8.2). The third section presents Luhmann's theory of communication (section 8.3). Then I lay out the communication theoretical perspective on social networks, first arguing that communication is routinely attributed to identities (section 8.4). This approach is confronted with the relational sociological formulations on storytelling (section 8.5). Section 8.6 proceeds from the construction of identities in communication to that of relationships. As a consequence, we can envision three kinds of interrelated networks: networks of recursively linked communicative events, social networks of actors in relationships, and cultural networks of symbols linked to each other (section 8.7). The last three sections offer extensions and applications of the perspective: the embeddedness of intercultural communication in network constellations (section 8.8), the construction of collective and corporate actors in communication (section 8.9), and a brief discussion of methods used to study networks in communication (section 8.10).

8.1. Forebears

The concept of communication has a long tradition in the social sciences. Apart from some casual usages, e.g., by interactionists, we find the first prominent conceptualization in the work of Claude Shannon and Warren Weaver (1949). They view communication as consisting of three parts: a sender, a receiver, and a channel of communication between them. The emphasis lies on the channel of communication that implies some sort of coding and decoding of the message. This is most obvious for the examples of mediated communication (telephone calls, letters, etc.). But it generally holds for all kind of communication: thoughts have to be translated into speech by the sender, and back into thoughts by the receiver.

Gregory Bateson and Paul Watzlawick provide a slightly different concept of communication that leads halfway to the model favored here. For them, communication not only entails the neutral transmission of some information. It also gives clues on how to interpret the communication. Bateson argues that gestures and facial expressions "frame" the situation in which communication occurs (1972, 178, 188, 215f). This "metacommunication" situates sender and recipient with regard to each other. It is more concerned with the relationships at play than with the information transmitted. In a similar vein, Watzlawick and his coauthors distinguish between the content level and the relationship level of communication. While the content level is about information, the relationship level

> refers to what sort of message it is to be taken as, and, therefore, ultimately to the *relationship* between the communicants. (Watzlawick, Beavin, and Jackson 1967, 52; original emphasis)

Consequently, social ties should not be seen as preexisting channels of communication. Rather, communication itself builds up, sustains, and changes social relations:

the messages constitute the relationship, and words like "dependency" are verbally coded descriptions of patterns immanent in the combination of exchanged messages. (Bateson 1972, 275)

Like Shannon and Weaver, Bateson, Watzlawick, and his co-authors conceptualize communication with regard to individual mental processes: the sender's intentions and the receiver's understanding. The development of social relationships would have to be modeled and studied with recourse to mental processes. However, their arguments concern the meaning inscribed in communication: apart from information about the world, it is accompanied by visual cues (gestures and facial expressions) concerning the meaning of communication, and it relates sender and receiver in particular ways.

8.2. Goffmanian interaction, conversation analysis, and Foucault's discourse

In addition, three diverse approaches offer important insights into communication without actually using the term: Goffman's dramaturgical approach, conversation analysis, and Foucault's discourse analysis all lean toward detaching communication from the subjective, as in Luhmann's later theory of communication.

In the first approach, the core of the *symbolic interactionist* tradition around George Herbert Mead and Herbert Blumer links social processes between people to their subjective processes ("internal dialogue"; see section 7.6). Erving Goffman and George McCall / J.L. Simmons focus more on the supra-personal dynamics of communication. Goffman points out that the "definition of the situation"—a core interactionist concept—is an interactive accomplishment, and that it does not even have to be shared subjectively:

The maintenance of this surface of agreement, this veneer of consensus, is facilitated by each participant concealing his own wants behind statements which assert values to which everyone present feels obliged to give lip service. ([1959] 1990, 20f)

People often hold up and follow a *social* definition of the situation as an "interactional modus vivendi" even if they *subjectively* disagree with it, for example, by tactfully saving each other's face from embarrassment (Goffman

[1955] 1967). In a similar vein, McCall and Simmons emphasize that individual identities—what people are seen as in their social environment—are the result of supra-personal processes ([1966] 1978).

In this perspective, social processes follow by and large the paths laid out in prior social processes. They construct individual identities and a socially agreed definition of the situation, saving both even against participants disagreeing or looking through staged performances. The inner thoughts of participants in interaction are less important than the communication relating the participants to each other. Indeed, Goffman later sketches interaction as consisting of a sequence of utterances "with each in the series carving out its own reference" (1981, 52, 72). Utterances construct an agreed-upon "participation framework" (1981, 124ff), building the ground for future utterances. The internal dispositions of actors only become important within the confines of communicatively established definitions of the situation.

In the second approach, *conversation analysis* takes a similar stance. It treats interaction sequences as "generic orders," looking for recurrent regularities of sequences of conversation. Any single utterance is treated as part of an interactive sequence rather than as the result of individual action (Schegloff 2007, xiv). Utterances lead to subsequent utterances. In line with Luhmann's formulations (to be discussed further), conversation consists of a "stream of communicated meaning." Utterances are analyzed with regard to their connections back and forth. The subjective meaning of the actors involved is ignored both in interpretation and in research practice.

This focus on communicative events and on the meaning in them also inheres in *Michel Foucault's discourse analysis*, the third approach. According to Foucault, a discourse consists of statements or utterances ("énoncés"; 1969, 41ff).[2] These discursive events follow systematic rules making for the orderliness of discourse. They obviously carry meaning by themselves since texts and statements are to be analyzed without recourse to their authors. Indeed, "authors" are projection points for Foucault to which we attach meaning and expectations on the basis of attributing utterances to them (1998, 221f).

---

[2] The English translation of Foucault's *The Archeology of Knowledge* coins these events "statements" ([1969] 1972, 28ff), while the French term "énoncé" is closer to the English "utterance." In the kinds of academic texts investigated by Foucault, most utterances are just statements. However, we think of a question, an order, an exclamation, or a declaration of love as an utterance, but not as a statement. Academic or intellectual statements typically only contain factual information and do not carry much of the kind of pragmatic aspect termed "Mitteilung" by Luhmann (which I translate as "message"; see further discussion).

Statements do not come from actors or authors; they emanate from their respective discourse, governed in their production by its rules.

Foucault's framework sees communicative events as parts of an orderly symbolic universe: a discourse. This parallels Luhmann's view that any given communication is the operation of a system that produces itself in these events ("autopoiesis"). In contrast, I take a more open perspective that does not necessarily see communication as part of larger ordered units—be they systems or discourses. But communication always picks up on previous communication—in the sense of building on "definitions of the situation" negotiated and established before.

Overall, these three very different approaches lead to a concept of communication that departs from the predominant focus on the subjective in sociological theories of action, practice, or interaction:

- Communicative events—and not only individual minds—*process meaning* in the sense of following the rules of discourse and offering a definition of the situation.
- Communication is *organized in chains or sequences of events*, with subsequent events building on the meaning established in previous events.

8.3. Luhmann's theory of communication

(a) Communication as self-referential

Niklas Luhmann picks up on these various approaches by focusing on observable processes between sender and receiver. He radically departs from the Shannon and Weaver and the Bateson and Watzlawick traditions by removing both the sender and the receiver from the concept. Luhmann conceptualizes communication as a *self-referential* process—an autonomous sequence of processing meaning (2002, 155ff). This perspective views utterances in a conversation not as based on subjective definitions of the situation or motivations. Rather, they spring from previous communication, as in Goffman's work, in conversation analysis, and in Foucault's discourse analysis. Any utterance has to pick up on the topics discussed, and it has to conform to the type of conversation taking place: a job interview, a scientific text, a discussion among colleagues, or a chat at a bar. It has to follow the expectations established in the course of previous communication (the so-cial definition of the situation).

Of course, communication would not be possible without individual minds ("psychic systems"). It actually breaks down without psychic systems participating (Luhmann 2002, 169ff). But Luhmann leaves this "participation" out of his basic model—partly for analytic simplicity, and partly because thoughts and other mental processes do not directly enter communication. When communication takes place, psychic systems operate relatively independently, pondering what has been said and what they want to say. But they also succumb to the general course of communication and continuously adapt to it. Conversely, psychic systems can have an impact on communication in the form of "irritations" that may or may not be picked up in the sequence of communication.

The disjunction of psychic and communicative processes is based on the inability of psychic systems to observe each other's operations directly, as Luhmann argues with reference to the figure of "the double contingency of action" by Talcott Parsons ([1968] 1977, 167f; Luhmann [1984] 1995, 103ff; see section 2.4). Action is contingent, according to Parsons, because participants in a social situation always want to make their action dependent on the actions of others. The double contingency enters with ego not only making her action contingent on alters' action, but also on alters' reaction to ego's action. In this situation, no action occurs because both parties (alter and ego) wait for the other party to act first. Parsons's solution to this problem is to assume a shared value system and common normative orientations of alter and ego. Luhmann, in contrast, claims that communication solves the problem step by step over time:

> At first, alter tentatively determines his behavior in a situation that is still unclear. He begins with a friendly glance, a gesture, a gift—and waits to see whether and how ego receives the proposed definition of the situation. In light of this beginning, every subsequent step is an action with a contingency-reducing, determining, effect. (Luhmann [1984] 1995, 104f)

Each communicative event reduces uncertainty by laying a track for future communication. It offers a specific definition of the situation that serves as a starting point for reactions (as in Goffman's account). Much small talk springs from this logic: people do not know what to expect or say to each other and engage in chitchat, avoiding difficult topics and keeping deeper

thoughts to themselves. Yet, something happens in this seemingly incon-
sequential chatter: a common ground develops, with a stock of knowledge
about the weather or about the neighbor's cat, and with a rough sense of
how the gossipers relate to each other. Luhmann's theory of communication makes the very unobservability of subjective dispositions part of its
model: communication operates self-referentially *because* psychic systems
can neither directly observe each other nor become part of the communication process.

> Assumption 8.1. Communication is a self-referential process. Every com-
> municative event picks up on the meaning processed, the definitions of the
> situation offered, in previous communicative events.

In a way, this does not really offer a new assumption for the theory to start
from, but a revised formulation of assumptions 7.1 and 7.2. Luhmann
builds his theory of autonomously operating systems (encounters, organizations, functional subsystems like the economy or law) on this model of
communication. I am skeptical about communication always falling into
these systems. But Luhmann's concept of social micro-events as communication provides a useful basis for a theoretical account of processes in
networks.

(b) Information, message, understanding

Luhmann conceptualizes communication as the "synthesis of three
selections, namely, information, utterance, and understanding (including
misunderstanding)" (1990, 3; [1984] 1995, 141ff; 2002, 157f). All of these se-
lect a possible state out of a wide range of options, or a particular meaning out
of a horizon of potential meanings. They combine to make for the meaning
of a communicative event (Figure 8.1).

- The *information* component resembles that of Shannon and Weaver
  and stands for Watzlawick and his coauthors' "content level" of communication. Information lies on the "factual dimension" of meaning and
  usually refers to something outside of the immediate situation of communication (external reference), for example, when people talk about
  the weather.

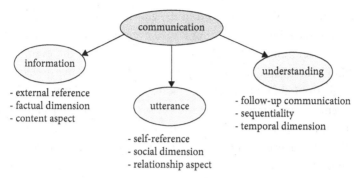

Figure 8.1. Three components/selections of communication according to N. Luhmann.

- The second selection is termed "Mitteilung" in German and translated as "utterance" by Luhmann. It includes the attribution of communicative events to actors, leading to questions concerning the "how" and "why" of communication (Luhmann 1990, 4). How and why do people talk about the weather? For the perspective offered here, this component can be slightly better captured as (relational) *message* than mere "utterance." What is the message conveyed from alter to ego (as opposed to the information) when talking about the weather? This message aspect of communication refers to the situation itself (self-reference), in particular to the social dimension of how the speaker or writer relates to the actors addressed. In this version, the "Mitteilung" or message parallels Watzlawick's "relationship component" and Bateson's "metacommunication" (see section 8.1).

- Finally, the *understanding* of communication establishes the meaning of the event in subsequent communication. This, of course, can involve misunderstanding of the subjectively intended meaning. The information and the message component are only decided upon in the course of their being understood. This third selection should not be seen as taking place on the mental, subjective level. Understanding is realized in the sequence of follow-up communication. For example, a yawn in a discussion about jazz music can be tactfully ignored (Goffman) and not affect the process of communication at all. Or it can lead to questions like "Are you bored?," which then could be taken as relating to the information aspect (alter finds it boring to talk about jazz music) or to the message aspect (alter is bored of ego).

Understanding is realized in the course of three subsequent turns, as the meaning of a communicative event is successively determined (Schneider 2000, 128). The first (say, the yawn) is the event in question. The subsequent event presents a tentative interpretation, e.g., with the follow-up question: "Don't you like jazz music?" The third event confirms this interpretation or rejects it by offering an alternative ("Uhm, no, I'm just tired."). From this point onward, the meaning of the original event is provisionally fixed. This changes something about future communication. For example, alter and ego may refrain from talking about jazz music. Every communicative event leads to expectations regarding the consequences of particular communication, and these expectations effectively structure the communication process.

Definition 8.1. Communication consists of three entangled components processed in social events: (1) the information component includes the factual content; (2) the message component consists of the attribution of the social event to an actor; (3) understanding determines the meaning of an event over the course of subsequent events.

(c) Communicative meaning

This basic model of communication meets the demands sketched at the beginning at this chapter (and developed in the last chapter). It avoids assumptions about unobservable subjective dispositions and processes and incorporates this very unobservability into its theoretical model. Communication as a process is placed on the social level—it focuses on the observable micro-events between actors. In contrast to the concepts of exchange and behavior, communication is firmly imbued with meaning. Communication unfolds meaning sequentially, step by step, as conversation analysis argues. We will see how networks as patterns of expectations arise in this process, stabilize, and change. But first, I discuss the implied concept of meaning.

Like Max Weber and Alfred Schütz, Luhmann conceptualizes meaning as the vast array of symbols and the connections between them (1990, 21ff; [1984] 1995, 59ff; see section 2.2). In contrast to the action theorists, Luhmann does not locate meaning only on the subjective level in the thoughts and dispositions of actors. The primary location of meaning is the social, with the mental heavily imprinted by the communicative processes of diffusion and attribution. Luhmann opts for the unfolding of meaning in communication as the main focus of sociological enquiry. But he concedes that mental processing (thoughts) is similarly based on meaning ([1984] 1995, 255ff; 1995b, 55ff). The overall model is not that different from that of symbolic interactionism with its simultaneous consideration of mental and intersubjective processes, but without assuming correspondence between the two levels (see section 7.6). While pointing to the importance of mental processes, Luhmann dismisses them from the concept of communication for analytical simplicity (and because of their unobservability).

However, he does see the dual processing of meaning in psychic and in social systems as a basis for their interplay through selective mutual observation and adaptation (Luhmann [1984] 1995, 210ff; 2002, 169ff). Thoughts adopt the symbols, schemes, and scripts processed in communication. And communication observes the irritations from the participation of psychic systems by attributing them to "persons" and attaching specific expectations to them (1995b, 142ff; see further discussion). All of this requires that we allow for communicative events (and not just thoughts) to carry meaning: a handshake *means* something, even if we do not know the intentions behind it. This meaning cannot exist independently of subjective interpretations (by sender and receiver, both of which remain unknown to others). But we can fruitfully consider meaning as inscribed in communicative events—and study how events relate meaningfully to previous or future events.

In a communication theoretical framework, *expectations* can also not be conceptualized as subjective. Rather, they form relatively stable structures inherent in the process of communication (Luhmann [1984] 1995, 96f). In Parsons's theory, action is always oriented to other people's expectations (Parsons et al. [1951] 1959, 19f). Properly modeled, we would have to say that individuals act on the basis of their perceptions of the (cognitive) expectations and prescriptions (normative expectations) of others (Galtung 1959; see section 7.3). Expectations can be more parsimoniously conceptualized as communicative patterns that are recurring but continuously renegotiated. These not only structure subsequent communication but also serve as effective points of orientation for psychic processes. Expectations are inherent in any communicated definition of the situation—they are the sediments of previous communication that channel future communication. In general, expectations form a "memory" of communication (Schmitt 2009). Similarly, Andrew Abbott views "social structures as the encoded memory of past process" (2001, 259; see also Abbott 2016, 11ff). Of course, we can think of communicative expectations as partly relying on mental states (and these on neurological structures, and on biophysical processes). But they can also be grounded in material symbols, for example, with a wedding ring or with documents. In any case, expectations have to be communicated (verbally, or through other signs) to be socially of consequence.

> Definition 8.2. The patterns of meaning established in previous communication, and pre-structuring communication in the present and future, are called expectations.

All social structures, not only networks, consist of such communicative expectations. A price of a product is an expectation that guides market behavior. The roles of professor or student are bundles of expectations defined in relation to each other, forming part of a larger array of expectations of a university (see section 5.2). Such expectations can concern factual questions (e.g., political issues or the quality of Italian food), but also the identities of those involved and the relations between them, social timing (e.g., an appointment or a conference schedule), and space (like territorial borders). Factual expectations result from the information component of communication (and structure it). The message component, in contrast, concerns the social dimension. Attributing a communicative event to an actor defines her/him/it, and it relates the actor to those addressed. This leads to the emergence of identities and of social networks (as I will argue in this chapter).[3]

8.4. Action in communication

So far, I have focused on the general theory of communication. How does this concept of communication lead to social relationships and networks? Luhmann was not really concerned with social constellations between the actors in communication. Therefore, I have to extend his theory of communication. In this section, I argue that communication yields factual expectations as well as identities of actors and relational expectations. The next section connects communication with the arguments about stories and relationships by White and Tilly.

---

[3] Talk about other people (in the information component) also affects expectations in the social dimension of meaning (Bergmann [1987] 1993; Raymond and Heritage 2006; Ellwardt et al. 2012). We can only roughly map the information component to the factual dimension and the message component to the social dimension of meaning.

Table 8.1. Two ideal-typical sequences of communication

Speaker	Sequence 1	Sequence 2
Alter	They're showing movie X at cinema Y tonight.	They're showing movie X at cinema Y tonight.
Ego	Uh, unfortunately, I don't have time. But let's get together next week.	Hmm, they've been showing it for a while now. Quite successful, it seems.
Alter	Yeah, good idea. What about Wednesday?	I read a positive review about it. C also said she liked it.

Let us examine a hypothetical example of two ideal-typical communication sequences. The situation features two coworkers conversing over lunch. Two alternatives start from the same initial sentence: "They're showing movie X at cinema Y tonight" (table 8.1). Then they proceed very differently. In the first sequence, ego responds that he does not have time tonight and suggests spending time together the week after. This reaction picks up on the relational message of the first sentence: it reads the sentence as the proposition to spend some time together off work. Alter's answer confirms this interpretation, but we do not know whether she initially meant it that way. Notwithstanding this contingency of subjective dispositions, the sequence establishes the meaning of the first sentence as a suggestion to go to the movies together. The information component (a particular movie will be shown at a specific cinema) is not taken up.

The second sequence arrives at an entirely different interpretation. Ego's reaction turns to the information component and leads to a conversation about the movie. Alter's response in the third turn still centers on the movie. This second sequence does not feature the interpretation of the first sentence as a suggestion to spend some time off work together. However, ego's response may also be read as deliberately avoiding the suggestion, and alter's final statement as an implicit renewal of that suggestion.

The two contrasting sequences thus accentuate the information and the message in the initial utterance. And they point to the ways in which the meaning of an event is negotiated and determined in communication. Two important conjectures result from this contrast:

(1) No matter what the initial intention behind a communicative event or its subjective understanding by a conversation partner,

its meaning will only be established in the process of communication. We do not know whether ego in sequence 2 understood the
initial utterance in the same way as in sequence 1. But we can see
that the first event comes to mean very different things in the two
sequences.

Conjecture 8.1. The meaning of a communicative event is determined in
subsequent events (in the "understanding" component), rather than by its
subjectively intended meaning.

(2) As in sequence 2, the subjective meaning behind communicative
events can remain uncertain for a while; it need not even ever be clar-
ified. Flirting is an example of conversation with the "real meaning"
left ambiguous. The different courses of the two sequences may reflect
a difference in subjective dispositions (ego's in particular), but we will
never know for sure. Also, the first utterance is more likely to trigger
sequence 1 if alter and ego already have a preestablished friendship.
Communication thus builds on the expectations from previous communication (in some cases called "friendship").

Conjecture 8.2. The meaning of a communicative event need not ever
be established unequivocally. Communicative events can offer ambiguous definitions of the situation, without follow-up events offering
disambiguation.

The two sequences also illustrate how the meaning structure of social
relationships and networks emerge in the course of communication. Over the
two sequences, alter and ego become visible as actors with specific dispositions
and with a particular relationship between them. Sequence 1 reads as more familiar and amicable, an example of two friends casually discussing their next
meeting. Sequence 2 shows us two colleagues talking about movies, but we
can suspect underlying tension out of an asymmetry in personal interest in
the relationship. A full interpretation of the two sequences obviously requires
considering its wider context. In any case, we see that every communicative
event leads to a definition of the situation comprising expectations about the
participation of the persons involved, and these expectations constitute their
relationship. How exactly does this happen?

Following Luhmann's model of communication, the human beings of alter and ego (or their psychic systems) do not themselves form part of the social process. But with the distinction between information and message, communication observes itself as *action* (Luhmann [1984] 1995, 164ff). Communication *operates* self-referentially as a sequence of communicative events. In the course of this self-referential process, every communicative event is *observed* (in subsequent communicative events) as being selected by somebody oriented to somebody else's behavior—the message component of a communicative event (see section 8.3). These observations concern irritations from psychic systems in the process of communication and tackle the question of what future irritations are to be expected from particular people in communication. Not that psychic processes could become part or even be observed directly in communication (or by other psychic systems). What people *think* remains fundamentally unknown—we only observe what they *say* and *do*. Communication thus becomes observed as a series of actions by the people involved. The message component leads to expectations in the social dimension of meaning about the participation of particular people in communication.

Conjecture 8.3. The message component of communication leads to expectations about the identities of actors, and about their participation in communication.

For example, at a scientific conference, the various presentations and contributions can (ideally) be seen as cumulating to a body of knowledge—a shared culture—consisting of factual expectations, e.g., about nanotechnology. Here the information components of many communicative events connect to each other. The message components, in contrast, do not lead to a body of knowledge. Particular presentations and contributions are seen as coming from particular scientists. Some of the contributions turn out to be more innovative, and others as dull or even flawed. Consequently, the scientists involved are observed as smart and innovative, or boring and outclassed. The message component leads to the attribution of capacities (e.g., smartness) and dispositions (e.g., ambitions) to the people involved. They are constructed as "persons" with particular qualities, with a specific cultural background (for example, from their home countries and institutions), and with interests and motivations (Luhmann 1995b, 142ff).

Both of these types of expectations are definitions of the situation and constitute different "memories" of communication—culture and identities.

In the same vein, sequence 2 from the example cumulates information components into a body of knowledge about movie X (where it is shown and when, that it is quite successful and positively reviewed, and that C liked it). Sequence 1 with the stress on the relational messages leads us to see alter and ego as persons with particular capacities and dispositions for action: alter is interested in going to the movies; both want to spend time off work together. The events in such sequences leave a trace, a "memory" of communication that subsequent events can build on and have to deal with.

8.5. Stories and identities

This observation of communication as action and the concurrent construction of the participants as actors with specific dispositions can easily be translated into the theoretical vocabulary of relational sociology (see chapter 2). Following Harrison White, social networks are "phenomenological realities as well as measurement constructs" (1992, 65; see also Ikegami 2005, 46f). The "reality" of a network is "phenomenological," based on meaning. This meaning of a network comprises identities that are linked by stories (1992, 66ff). Unfortunately, White's concepts of identity and stories are a little fuzzy. *Identity* encompasses both psychic systems as the sources of irritation in communication, and the communicative observation of these irritations as "persons" (2008, 10f). Let me disentangle these two senses of identity as (1) primordial sources of action and as (2) social constructions and stick to the second. According to White:

> Identity here does not mean the common-sense notion of self, nor does it mean presupposing consciousness and integration or presupposing personality. Rather, identity is any source of action not explicable from biophysical realities, and to which observers can attribute meaning. An employer, a community, a crowd, oneself, all may be identities. (1992, 6)

In this constructivist stance, identity derives precisely from the "attribution of meaning" as a "source of action." The attribution of meaning takes place in the form of *stories* that mark an identity as different from something else, relating it to other identities. White sees such stories as emerging in the course of communicative events ("switchings"; White 2008, 12, 27ff; Godart and White 2010; see section 7.8). They are made of verbal and non-verbal meanings that characterize relationships (as different from other relationships) and that define the identities involved in relation to each other.

Charles Tilly elaborates on how stories define and relate identities. His "standard stories" (the prevalent type of narrative in informal conversation) portray individuals as responsible for events with their capacities and dispositions (2002, 8f, 26f; 2006, 64ff). Structures and unintended consequences of action are little considered. In these stories, the social world consists of autonomously acting individuals:

> To construct a standard story, start with a limited number of interacting characters, individual or collective. [. . .] Treat your characters as independent, conscious, and self-motivated. Make all their significant actions occur as consequences of their own deliberations or impulses. (Tilly 2002, 26)

The identities of actors arise in these stories through the attribution of actions to them (Tilly 2002, 80), as in Luhmann's account of communication. Behind these actions, motivations or intentions are inferred that make identities act the way they do (Blum and McHugh 1971). Action is constructed as rooted in the subjective meaning that the actor holds of a situation. Our everyday communication thus adopts an action-theoretical stance: Communication regularly attributes incidents to actors bringing them about, and underlying motivations (beliefs and interests) to the actors. From the perspective adopted here, motivations appear not as subjective reasons for behavior, but as typical patterns for how to "understand" behavior in communication (as in C. Wright Mills's "vocabularies of motive"; 1940).

> Definition 8.3. Stories recount social events by attributing them to actors, with subjective dispositions and capacities for action as driving the events.

With these elaborations, the story concept translates well into the framework of *expectations*. If actors are constructed as acting out of subjective dispositions, these attributions lead to expectations about the behavior of the actors in a new situation. For example, it is expected that a person's liking or disliking of jazz remains stable from one situation to the next. Similarly with capacities for action: a scientist who was observed to give a good presentation at one conference is expected to be similarly smart and witty at the next. For these expectations to arise, communication first has to be attributed to actors. Second, as in Tilly's standard stories, the capacities and the subjective dispositions of the actors have to be seen as *causing* the yawn in the discussion about jazz music or the good presentation at the conference. And this sets the course for future communication, for example, when talk about jazz is avoided, or a scientist is invited to the next conference.

The basic linguistic structures are geared to an action-theoretical description of social processes, with "subject, verb, and (sometimes) object" (Tilly 2006, 65). This renders communication-theoretical accounts difficult. Here, communicative events and not actors "do" things; actors are "constructed" as the sources of actions; interests and motivations are attributed, rather than objectively causing things to happen. This poses a serious obstacle: theoretical accounts in the social sciences also have to convince as "stories." Since the communication-theoretical perspective does not conform to standard storytelling, it is equally difficult to write and to grasp. This book, too, often resorts to standard linguistic forms (of "people doing something") in order to avoid awkward and overly technical language.

8.6. Identities in relationships

These considerations fit nicely with Luhmann's account of communication. Communication is routinely observed as action with the message component. This leads to the construction of actors with specific dispositions for action (based on the motivations we attribute to them in standard stories), and to expectations about their future behavior. But up to now, we only see isolated individuals with their dispositions and autonomous actions. How are these individual identities related to each other to form a network? In this regard, the message component of communication has to be interpreted as not only pointing to an identity as the source of an utterance, but as inherently relational.

Looking back at the two exemplary communication sequences in table 8.1, we find something more at play than the construction of isolated identities. In sequence 1, alter not only appears as somebody who likes to go to the movies. Over the three-step sequence, the initial sentence gets to signify a suggestion to go to the movies together, maybe out of a wish for friendship with ego or even a romantic interest. Ego's response is not only about his available time but also reciprocates alter's amicable or romantic interest. From sequence 2, we learn something about alter's interest in movies. But we also suspect alter to be interested in shared activities with ego, and ego as probably avoiding alter's advances. In both instances, the message component results in expectations about the individual actors involved, and about their relation to each other, as captured by Watzlawick's concept "relationship level."

For this, communication is observed as action, but in particular as "social action" in the sense of Weber. When understanding picks up on the message component, it not only asks for individual dispositions but also for the orientation of the underlying subjective meaning to the behavior of others. What does ego mean to alter? Does alter hope for a joint evening activity? Does she want a deeper friendship (or maybe she already thinks she is friends with ego) or only to reduce existing tensions with a work colleague? These questions are a routine part of all communication but rarely explicit or answered explicitly. With the message component, communication incessantly looks for clues about the relationship between the participants. Which propositions will be successful, and which ones will not? From these routine observations, *relational expectations* result about the ties between the participants. Identities are related and defined in relation to each other, thus connecting to a network.

> Conjecture 8.4. Communication (in its message component) is routinely observed with regard to its implications for the relationships between the actor speaking and that or those addressed. This leads to relational expectations structuring future communication.

Therefore, the story concept has to be spelled out in more detail. In network theory, stories are not about isolated identities. Rather, stories place identities in relation to each other. The subjective meaning of actors is still part of the story—but primarily in its orientation to others. We are dealing here with admiration, enmity, trust, mistrust, or envy. These subjective orientations do not feature explicitly in communication, but they become part of the stories told about identities and their relationships with each other in the social dimension of meaning. The message component more often than not remains unspoken. Nevertheless, it carries important consequences by establishing expectations about how particular actors will behave to particular others.

But is the attribution of social events to actors always spelled out in the form of stories? In the two exemplary sequences discussed in this chapter, communicative events lead to expectations, without stories actually told about these events. Stories lead to expectations, but expectations can also develop without stories. However, the fundamental process requires attribution, *as in* stories. We can conjecture that communicative events are monitored routinely, but implicitly, with regard to vocabularies of motive (Mills) underlying them. Does a communicative event (such as the one triggering the two exemplary sequences) signify amicable or romantic interest?

Many relationships are constructed and negotiated by adherence to typical repertoires of action, which Viviana Zelizer terms "relational work" (2005; Bandelj 2012; see section 2.7). Relationships are demarcated through the communicative events taking place in them. A kiss is surely out of line among strangers, or in a friendship. Even an attempted kiss will lead to a recalibrating of relational expectations. Stories told about the communication between actors are surely a part of the definition of their relationship. But we can arrive at such a definition, or renegotiate it, simply through communicative practices. Tilly's main concern lies with the "political identities" of collective mobilization, parties, and governments in political discourse, and with the relations between them (2002). These rely on stories told about them, more than individuals and personal relationships. But the general mechanism does not seem to require stories, only the attribution of events to actors (and the inferring of underlying relational motivations).

From this perspective, social networks do not consist of individual nodes that are linked to each other. Rather, the links are the primary feature of networks. As I have argued, social networks feature two levels: (1) regularities in patterns of communication, and (2) relational expectations underlying these regularities (the "meaning structure of social networks"; see section 2.1). Relational expectations develop through the course of communicative events out of their attribution to actors, and they make for predictability and regularity in communication.

Individual nodes themselves do not feature in these two aspects of the network, at least as long as we think of them as human beings of flesh, blood, and thought. Neither cognition nor digestion nor blood circulation becomes part of social networks. But they can be communicated about, and we attach expectations to them. Human beings are part of the network's environment, and as such they are continuously observed by the communication in the network—by attributing individual motives behind specific observable behavior. The nodes connected in relational expectations are not human beings but their observation in the form of *persons*, that is, the bundle of expectations attached to a human being structuring her participation in communication (see earlier discussion). In this sense—as constructed actors who are expected to behave in specific ways in relation to other actors—individuals (but also organizations or collective identities) play a vital role in networks: through these actors, relationships are linked to one another to form a network—as the term "node" implies (see section 2.6).

How can we tell whether a relationship between two actors exists? What is a social relationship? In the communication-theoretical perspective, social relationships consist of relational expectations between the actors involved, reflecting regularities in the process of communication, and bringing them about (see definition 2.4). It is not only important how often you meet somebody, say, the waiter in your favorite café (frequently), or your best friend from school (almost never). It also matters how you communicate once you meet. Communication often jumps directly into an intimate and amicable mode with long-time friends and siblings, even after extensive periods without communication. In contrast, the communication with the waiter may remain relatively formal and superficial, even when meeting regularly. But of course, you could develop a friendship with the waiter in your frequent encounters. A relationship comprises the idiosyncratic expectations concerning the behavior of two actors in relation to one another, insofar as they differ from general expectations about the behavior of people belonging to particular categories (e.g., male/female) or occupying formal roles (e.g., waiter/customer). As soon as the communication between two people deviates from general or purely formal courses, such idiosyncratic relational expectations develop.

> Conjecture 8.5. Relational expectations map and prescribe the communication between the actors involved. This makes for observable regularities of communication.

As I argue in section 2.7, communication draws on culturally available relationship frames (love, friendship, etc.) to structure itself. This makes for the relative independence of relational expectations and regularities of communication (which are more affected by opportunities for contact), even though both are tightly connected. Also, the adoption of relationship frames leads to the clustering of different ways of communicating in "types of tie" in terms of their "action profiles" (Martin 2009, 11, see section 2.4).

Network research bases its formal analyses on whether particular ties exist or not. I stipulate that that question is often impossible to answer. Rather, we should ask: What kind of relationship do we find between two actors at a certain point of time? What kinds of expectations govern their communication? Of course, this more qualitative assessment has to be reduced to ties (1) and non-ties (0) for network analytic algorithms. But this quantification is always imperfect. We continuously have to ask ourselves about the "reality" of ties and non-ties. And we have to distinguish between different types of tie, to avoid conflating them without attention to the meanings connected to them, and to the kind of communication taking place in them. Finally, we have to pay attention to the dynamics of communication in relationships. The relational expectations between two actors continuously change, and this means that a tie measured on one day might look very different the week after.

8.7. Three kinds of networks

All of this suggests that we are dealing not with one, but with three types of networks:

(a) Communication itself is organized as a network, with micro-events (as the nodes) related to each other. As operations, communicative events always connect to immediately preceding events and realize their meaning only in the follow-up sequence of communication. But on the level of observation, communicative events can relate back to a wide range of past events, and they can anticipate future events, such as responses or planned events. However, *communication networks* are mainly organized recursively, because it is much easier and more common to refer to actual communicative events in the past than to anticipated events in the future. For example, if we look at scientific communication, texts relate to other texts already published, leading to a recursively organized citation network of texts (Leydesdorff 2007).

(b) From this network of communication, the *network of actors* emerges through the attribution of communicative events as actions of identities. If we identify authors of texts, we can interpret the recursively organized citation network as the ongoing dialogue of scientists in a field. Actors are connected to each other through relational expectations. I only term this network of actors "*social network*" because only this one is governed by the basic constellation of double contingency between actors and by the gradual build-up of relational expectations (see section 2.4).

(c) The third network consists of the *structures of meaning* available within a given context, its "culture" (Mohr 1998; DiMaggio 2011; Fuhse et al. 2020). As argued in section 2.2, meaning is organized by linking symbols to each other. Terms like "chair," "table," and "sit" only carry meaning in relation to each other. In this sense, the body of knowledge emerging in scientific discourse can be mapped in networks of concepts (Fuchs 2001a; Leydesdorff 2007). The level of culture most relevant in the study of social networks consists of the codes of conduct, the relationship frames, and the social categories that are used in communication to relate actors (see sections 2.1 and 5.9). Relationship frames like friendship and love can acquire different meanings in separate network contexts and lead to very different network structures (Yeung 2005). Similarly, the meanings of role categories or of social boundaries like gender or ethnic descent are important for the structuring of social relations (chapters 4 and 6). They differ in the expectations connected to them by cultural context, and they change over time.

These three networks—of communicative events relating back and forth, of actors connected in expectations, and of symbols and terms related in culture—are layered and interconnect in a number of ways. However, their logics of structuration and their dynamics differ greatly. I cannot offer a full account of their relations to one another. Instead, the remainder of this chapter offers two areas of implication and application of the communication-theoretical perspective: intercultural communication in networks (section 8.8) and networks of collective and corporate actors (section 8.9). And it briefly discusses methods to study social networks in the process of communication (section 8.10).

8.8. Intercultural communication

A framework that combines communication, meaning, and social networks is particularly suited for the study of intercultural communication in four regards: (1) It allows taking social context into account; (2) it points to particular problems of intercultural communication and their consequences; (3) network-based techniques for successful intercultural communication can be identified; (4) it focuses on the consequences of intercultural communication for the development of relational expectations in social networks (Fuhse 2014). I briefly discuss these four points here.

(1) Densely meshed networks—for example, in friendship groups, in classes or milieus, in ethnic groups, or within nation-states and academic disciplines—lead to the emergence of particular cultures (see section 3.3). In particular, they develop differing moral standards for the attribution and the evaluation of social action. Intercultural communication has to bridge these different symbolic repertoires. Following the arguments of Ronald Burt (2004) and of Mark Pachucki and Ronald Breiger (2010), structural holes between network clusters tend to be *cultural holes*. The "weak ties" between them come with specific advantages for the individuals occupying the brokerage positions. But they are also prone to the problems of intercultural communications: the differing symbolic repertoires between academic disciplines, between ethnic groups and the generations, or between national cultures lend themselves to misunderstandings across the divide.

Definition 8.4. Intercultural communication takes place between densely connected networks of communication with their distinct cultural codes.

(2) These misunderstandings are particularly severe if they pertain to the message component of communication. Differences in the interpretation of the information component vanish by way of proper translation, but at least they are easier dealt with than in the message component. As the previous example of two communication sequences shows (section 8.4), the message component refers to relational dispositions of the actors involved that often remain implicit. The proposition to go to the movies together was not verbalized but inferred and realized in subsequent communication. If we only look at the information component, sequence 1 (which picks up on the message component) does not offer any meaningful connection between the first and the second turn. Gregory Bateson points out that

relational communication (which he terms "meta-communication") is often based on gestures and mimic (1972, 178, 215f). Words are easy to translate, nonverbal communication is not.

The interpretation of the message component rests on typical "vocabularies of motive" (Mills 1940). Where vocabularies of motive differ, potentially self-reinforcing cultural misunderstandings ensue (Gumperz 1982, 153ff). For example, French conversation is typically characterized by interruptions. These are meant to convey the message: I find it interesting what you say and therefore like to engage in discussion. In Germany, interruptions are seen as signifying lacking interest: if you cared about what I said, you would let me finish my sentence. Franco-German conversations thus often leave both conversation partners frustrated. The French tend to interrupt the Germans to display their interest, while the Germans let the French finish—and both have the impression that the other party is not really interested in what they are saying (Lüsebrink 2008, 32).

Conjecture 8.6. Misunderstandings in intercultural communication tend do concern the message component, and the interpretation of relational meaning.

The consequences of these message-related misunderstandings concern the development of relational expectations (and less that of a shared body of knowledge). Business partners in international joint ventures will have fewer problems agreeing on information than arriving at a definition of their identities and of the relationship between them. Social networks across cultural differences remain more fragile and insecure in their expectations than those within homogeneous cultures. They tend to acquire less predictability and regularity than where communication can draw on common schemas of interpretation.

(3) The most common technique for dealing with cultural differences is the classification of individuals in conjunction with stereotyping their behavior. Categories like "Chinese," "women," "working class," or "youth" offer convenient shortcuts for the interpretation of differences in behavior. Of course, the stereotypes attached to these categories cannot do justice to the widely differing action scripts among the

people indiscriminately called "Chinese," for instance (between rural and urban areas, between North and South, between Mainland China and Taiwan, between young and old etc.). But it may be even more misleading to avoid stereotypes altogether. When confronted with other people, we rarely know much about their idiosyncratic cultural imprint. Some ideas about a typical Chinese vocabulary of motive will probably work better than none at all. Rather than avoid stereotypes, it is advisable to start from basic stereotypical assumptions, to revise them carefully in the course of communication, and to make the cultural differences the topic of discussion in a sort of "exploratory communication." Stereotypes will thus enter the information component, rather than remain confined to the message component—which is probably a less dangerous way of dealing with them.

Another common technique is the use of mediators in intercultural communication. A mediator enlarges the basic dyad of communication to a triad. Such mediators usually know both cultural backgrounds and the ensuing cultural misunderstandings well. In addition, on the structural level, the mediator introduces "closure" (in the sense of James Coleman, 1988) into communication. Alter and ego are now indirectly connected through a third party, which increases the level of social control and facilitates the emergence of a trust relation. If both are already previously related to the mediator, we expect transitivity to kick in (Cartwright and Harary 1956): the strong ties of both alter and ego with the mediator make the development of a cooperative relationship between them more likely. Ideally, the mediator becomes superfluous as the relationship between alter and ego gradually acquires a cooperative history of its own.

(4) If networks as structures of relational expectations result from communication, intercultural communication, too, should lead to the emergence, stabilization, and change of social networks. This can lead to two ideal-typical instances of network constellations, one the result of "failing," and the other of "successful" intercultural communication.[4]

---

[4] Luhmann would argue that communication does not fail as long as it does not stop—and both constellations would count as successful crystallizations of patterns of communication. Thus, the terms "failing" and "successful" are not meaningful on the level of communication theory, only from the standpoint of an outside observer evaluating specific ways of communication and specific sociocultural constellations as preferable.

Intercultural communication that is laden with misunderstandings, contradictions, and conflict tends to affirm existing stereotypes or even develop new ones. In the sense of Gregory Bateson, a "double bind" (1972, 206ff, 271ff) holds both parties captive. The other's behavior confirms one's own rejective stance, and vice versa. As in the study of conflict in an English suburb by Norbert Elias and John Scotson ([1965] 1994), social boundaries and cultural differences self-reinforce (see section 4.1).

Successful intercultural communication, in contrast, leads to the development of collaborative social relationships across the socio-cultural divide. Structurally, a "bridge" forms across the hole in the network. But also the cultural forms from the two contexts come into contact. This can lead to the emergence of Creole linguistic forms and hybrid cultures of transition (Hannerz 1996, 66ff). Just as structural holes yield opportunities for mediation and the privileged access to information, cultural holes come with the possibility of blending cultural elements from both network contexts creatively (Pachucki and Breiger 2010). This can result in the development of a new style, as in Collins' study of the emergence of new schools of thought in the history of philosophy (1998; White 1993, 72ff).

Conjecture 8.7. Successful intercultural communication leads to the formation of social relationships across the cultural divide, and to the combination of cultural forms. Failed intercultural communication tends to reinforce structural separation and cultural stereotypes.

Overall, the communication-theoretical perspective on social networks allows for the analysis not only of the conditions of intercultural communication, but also of its consequences. The wide concept of intercultural communication deployed here can focus on very different social phenomena: the contact between different schools and disciplines in science and academia; between ethnic and lifestyle groups at work, in school, or on vacation; and also between women and men (Tannen [1990] 2007, 42f, 47, 281f). In all of these cases, the process and the results of intercultural communication depend on the network constellation and on the communication techniques used. Consequentially, old lines of difference and conflict are reproduced, or new sociocultural formations emerge.

8.9. Collective and corporate actors

For reasons of simplicity and intuitive accessibility, I have focused on networks of expectations between human individuals so far. Can we analyze and interpret networks between collective and corporate actors in the same way as those between individuals (Laumann 1979)? The perspective adopted here requires that something is seen as an acting entity, a "source of action" (White) in the network. This constructivist definition of actors leads to the attribution of agency as the decisive variable for something to count as a node in a social network (Fuchs 2001b). The entity in question (be it a person, a collective or corporate actor, or even transcendental identities like gods or spirits) has to be regarded as reasonably *coherent*, that is, not composed of separate parts following their own will, and as acting out of *inner dispositions*, that is, not determined by outside forces. Accordingly, a traffic jam would not count as an actor because of being composed of different parts, but a protest march would because of acting coordinately and pursuing a common objective. As a result, relational expectations will be attached to the protest march regarding its behavior vis-à-vis political actors, the police, and bystanders.

> Assumption 8.2. Relational expectations can concern individual, collective, and corporate actors. They result from the attribution of behavior and of underlying dispositions of action (toward other actors).

Let us look more closely at social movements as an important case of collective actors. In contrast to individual actors, their unity is not immediately obvious. Rather, it is realized in the course of communication and in successful storytelling about the movement (Tilly 2002). In such storytelling, very different behavior is portrayed as the coordinated action of the collective. Symbols and slogans flag forms of protest as done by the collective and not as isolated individual acts. These symbols construct an identity in the diversity of individuals and their behaviors, and they draw a boundary to the outside (of opponents and audience; see section 3.5). This symbolic identity and this boundary become a projection point both inside the collective and in its relations to the outside world—to other collective and corporate actors, as well as to the public.

The collective or corporate actor is then defined by the expectations about how it will behave—including to other collective and corporate actors—but also by the behaviors expected of its members (on behalf of the collective or corporate actor). Similarly, individuals are defined by what we expect of them, based on their past behavior. Of course, the collective identity constructed internally (self-ascription) and that projected to the outside world (other-description) do not necessarily match. A terrorist group sees itself as fighting for a good cause by all means necessary, whereas much of the outside world regards it as a crusade against humanity. But the two will interplay, with the relations to other collectives or corporate actors becoming a part of the internal construction of collective or corporate identity, and vice versa.

Generally, we have institutionalized models for what an actor looks like and what she, he, or it does (Meyer and Jepperson 2000). These models change over time and from one cultural context to another. But at any point in cultural time and space, they provide clear frames for the construction of social actors, and for the interpretation of their actions (see section 5.9). For example, the (changing) repertoire of social movements can be seen as marking a relatively new class of social actors, as scripts for the construction and the identification of social movements as collective actors (Tilly and Wood 2009).

As a consequence of the attribution of actions to the collective, expectations about future behavior develop. The construction of values and interests (the "motives" of collective actors) in a communicative storytelling leads to expectations about the position of social movements in the ensemble of political and civil society actors. To take another example, the identity of a gang (its position in the network) is very much a result of previous interaction (e.g., killings) with other gangs—and this network position affects what homicides take place in the future (Papachristos 2009). As between individuals, the relational expectations in the network of gangs give orientation and make specific communication more likely and other communication improbable. The communicative dynamics of networks of collective and corporate actors does not differ in principle from those between individual actors.

However, collective and corporate actors are not natural entities, or observed as such. Their attribution of actorhood relies on the observation of internal coherence in its behavior. I illustrated this earlier with the lack of coherence of a traffic jam, in contrast to the relatively coordinated behavior of a protest march. Collective and corporate actors are assigned actorhood, with the construction of expectations regarding their behavior, to the extent that they seem to be acting in unison. Tilly termed this the WUNC principle: social movements portray themselves as "worthy, united, numerous, and committed" (2002, 120; Tilly and Wood 2009, 4, 19). The perceived worthiness of social movement actors may differ. But their unity, numbers, and commitment depend on the cohesion of interpersonal networks among their members. The more that positive social relationships cluster within collective actors, the more convincing their construction of collective identity and the higher the normative pressure to take part in collective behavior (Gould 1995). Also, social relationships are key for the recruitment of new members (McAdam 1986; Opp and Gern 1993). Conversely, if the collective is not recognized as relevant (or as acting in unison), fragmentation and dwindling numbers may ensue. We can hypothesize an interplay between communication inside the collective and the communication in the network with other collective and corporate actors.[5]

> Conjecture 8.8. The communication and network structures within a collective actor and the communication in the network with other collective or corporate actors (recognition, conflict, alliances) affect each other. Collective actors with more social cohesion and mobilization tend to be better capable of coordinated action and more relevant in the overarching network.

We can thus construct a model with multiple levels of networks—e.g., between individuals within social movements and political parties, between those political actors in national political discourse, and between nation-states on the international level. These levels will not be independent of each other. Social movements and political parties need interpersonal networks for the recruitment and commitment of its members. And nation-states have to be capable of coordinated action (unlike failed states) to be relevant in international relations.

8.10. Excursus on methods

What are the implications of the theoretical considerations in this chapter for the study of social networks? The substantive expectations concerning the interplay of social networks and meaning, as outlined in chapters 2–6, are not much affected. They are by and large compatible with theoretical

---

[5] Corporate actors are more complex because they rely on formal rules and often payments for membership to ensure the commitment of their members and their capacity to act (Kühl 2020).

perspectives based on the notions of action, interaction, exchange, transaction, and communication. The communication theoretical approach instead suggests a different way of studying networks and the processes in them. Instead of focusing on the individual perception of network ties and underlying subjective dispositions, research could investigate social relationships as regularities of communication and as developing, solidifying, and changing in sequences of communicative events. This leads away from sociometric surveys toward either in-depth qualitative analysis of communication or large-scale non-reactive examination of events as in computational social science.

Of course, we can interpret the subjective reports on networks in interviews as reflecting communicative patterns. Many survey questions, such as the Burt name generator, ask individuals about communication, rather than preferences or feelings:

Looking back over the last six months [. . .] who are the people with whom you have discussed an important personal matter? (Burt 1984, 331)

Also, respondents probably assess communication with others when applying summary labels like "best friends" or "people important to you." However, they use very different subjective yardsticks. Consequently, a lot of them give diverging answers, e.g., when people disagree on whether they are friends with each other. The resulting "asymmetric friendships" are hard to interpret (Friedkin 1980, 413). This poses serious problems of unreliable measurement. Also, these survey methods assume and suggest network ties to be stable entities that either exist (1) or not (0), in contrast to my arguments earlier (section 8.6).

Of course, I do not want to declare the extraordinarily successful research tradition building on these methods as fundamentally mistaken. In particular, the study of egocentric networks in surveys cannot well do without name generators and dichotomization. However, the communication-theoretical perspective proposes complementing the predominant network surveys with methods that examine communicative events directly and non-reactively. There are two basic ways of doing so:

(1) We can detect social relationships based on the *regularities of communicative events*, in line with assumption 2.1 and conjecture 8.5. This is frequently done in studies of computational social science that analyze

large datasets of non-reactive communication (Kitts 2014, 281ff; Kitts and Quintane 2020, 83f). Events between actors, and making for their ties, include emails (Kossinets and Watts 2009), literary reviews (de Nooy 2011), contact initiation and responses in online dating (Lewis 2013), and the forwarding of patients among hospitals (Kitts et al. 2017). Practically all kinds of social events can be considered "communication" and part of relationships. This approach faces three challenges:

- It is not clear *how many events* constitute a tie, or how *regular* they have to be. For instance, a single citation probably does not make for a "connection" between scientists, just as one email should not count as a "relationship" between employees of a company or students in a university. The nature of the data suggests gradual measures like "frequency." These lie closer to the gradual understanding of relationships advocated in this book than the 0/1 logic of network questionnaires. But they do not square well with the formal network analyses relying on graph theory.

- Most social relationships are not characterized by one particular kind of communicative event, but by the combination of *different kinds of communication* (Hinde 1976; Martin 2009, 11; see section 2.4). For instance, emails between company employees indicate friendship when combined with invitations to birthday parties and gossiping, but they could also form part of a strained work relationship. In this vein, Wouter de Nooy and Nils Kleinnijenhuis analyze two types of communicative events in election campaigns: attack and support (2013). Quantitative text analysis allows investigating communication in much more detail: in our study of parliamentary interaction in the Weimar Republic, we discern valued ties (from support to attack) between parties based on 14 typical kinds of interruptions and reactions to speeches (Fuhse et al. 2020, 11ff).

- We have to ask about the *meaning of communicative events* as well as that of social relationships. How exactly are two scientists related if one cites the other? What does it mean for one political party to interrupt the speeches from another with laughter? These questions do not pertain to the meaning of a communicative event per se, but only with regard to its implications for the relation between two actors. It might be possible to detect these meanings partly from quantitative analysis, by examining what kinds of events feature systematically together in ties (Gibson 2005; Fuhse et al. 2020, 10ff). However, it seems prudent to combine this coarse structural approach with the qualitative interpretation of communication, just as network generators are subjected to qualitative validation (Bailey and Marsden 1999).

(2) This in-depth *qualitative interpretation of communication* in networks complements the formal approach in (1). Building on Erving Goffman, we can trace how definitions of the situation, including identities and relationships, are activated and negotiated in communication. For this, we have to examine events like citations, interruptions in discussions, and particular linguistic forms for their relational implications. Ideally, this leads to the identification of typical forms of relating—types of relational events—that can then be subjected to the formal methods mentioned earlier. To do so, we can draw on the methods and studies from conversation analysis (Maynard and Zimmerman 1984; Goodwin 1990; Lerner 1996; Heritage and Raymond 2005) and from interactional sociolinguistics (Brown and Levinson [1978] 1987; Schiffrin 1994, 97ff; Spencer-Oatey 2002; Locher and Watts 2005). These already give us important insights into how linguistic forms (e.g., discourse markers like "oh" and "y'know"; Schiffrin 1987) and types of communicative events (e.g., yes/no questions; Raymond 2003) relate actors in a participation framework. Following the arguments in this chapter, the analysis has to focus on the "message" aspect of communication (its "pragmatics") rather than its informational content. Generally, sociolinguistics is yet rarely appreciated in its potential contributions to the qualitative study of communication in social networks (Diehl 2019).

Communication does not only indicate social relationships by following the expectations inscribed in them. It can also bring them about or change them. Instead of looking for recurrent events in ties, the sequential nature of communication data allows us to investigate interactional dynamics: What kinds of events lead to what kinds of follow-up events, slowly and gradually making for the formation of ties, or for their change? This temporal perspective is well established in the qualitative approaches to communication. Conversation analysis, in particular, examines sequential patterns of turn-taking and of adjacency pairs of communication events typically following each other, like questions and answers (Schegloff and Sacks [1973] 1984;

Sacks, Schegloff, and Jefferson 1974; Schegloff 2007). Conversation analysis looks mostly for general patterns in the order of types of events. Our approach would build on this to discern how particular events trigger specific responses and build up to typical network constellations, such as hierarchical transitivity (Chase 1980).

Quantitatively, the dynamics of events in networks are investigated in relational event models (Butts 2008; DuBois et al. 2013) and in network autoregression models (de Nooy 2011; de Nooy and Kleinnijenhuis 2013). Here, the past events between two actors (repetition, reciprocity) and in adjacent ties (transitivity) as well as structural factors are used to predict whether a particular communicative event occurs in a tie. As in conversation analysis and interactional sociolinguistics, these studies zoom in on the sequential nature of communication. The challenge, then, is to consider communicative events not on their own, but as nested in social relationships. Events can be predicted from an existing network constellation in a straightforward way (McFarland 2001). But events do not only follow preexisting patterns of expectations: they bring social relationships about and change them (McFarland, Jurafsky, and Rawlings 2013). To my knowledge, no formal models have yet managed to disentangle relatively inert social relationships and short-term dynamics of interaction.

Another challenge concerns the variety of communication in relationships: How can we investigate different kinds of events—e.g., citation, collaboration, coauthorship, review, and invitation for a talk among academics—and their sequential repercussions for each other? The sequential ordering of events in networks already makes for three-dimensional data matrices (actor × actor × time). Now we also have to examine these three-dimensional data matrices for multiple kinds of events and their potential cross-effects in time.

Overall, the methods discussed here dissolve static vision of networks as stable structures. Instead, they draw attention to the histories of events in networks. This focus on communication rather than actors befits the time-stamped process-generated data from archival sources or from the Web 2.0, as often studied in computational social science and digital humanities (de Nooy 2015; Kitts and Quintane 2020). The key challenges here are the entanglement of relatively stable social relationships and short-term dynamics and the multiplicity of communicative events in social networks. In addition to the formal advances, network analysis would do well to draw on the qualitative methods and insights on the temporal order and the relational implications of linguistic forms and types of communicative events from conversation analysis and interactional sociolinguistics.

8.11. Conclusion

The communication-theoretical perspective on social networks draws on a number of sources. It centrally builds on Niklas Luhmann's concept of communication and his account of meaning, arguing that social networks (as structures of meaning) evolve in a *sequence of communicative events* recursively linked to each other. However, I do not embrace Luhmann's wider theory of autopoietic systems. I focus instead on networks of relationships between actors as *one* structure of expectations emerging in and guiding communication. This leads to relational sociology, with Harrison White as its central theoretical figure. I adopt White's assertion that social networks are phenomenological realities, linking *identities* through *stories*. But social relationships should not be read as mere stories. Relationships consist of *relational expectations* about the behavior of actors in relation to other actors. These can be spelled out in stories but do not have to be. In stories and in communication in general, communicative events are attributed to actors (Tilly). This attribution leads to the projection of inner dispositions and to expectations about their future behavior in relation to other actors. Here, Paul Watzlawick and his co-authors' distinction between the content aspect and the relationship aspect of communication is relevant, as are Erving Goffman's insights about the negotiation of definitions of the situation in interaction.

This theoretical inventory allows me to answer the four questions from the start of this chapter:

1. What exactly are social networks, and where do they come from?

    Social networks are not mere patterns of links, but structures of relational expectations between actors. This "meaning structure" emerges in the process communication from the attribution of communicative events and underlying dispositions to actors. In turn, it guides the future course of communication, making for recognizable patterns.

2. How do networks change, and why should they be relatively stable?

Networks of expectations are continuously confirmed or modified in the course of communication. But since communication builds on pre-existing networks, it tends to conform to these expectations, confirming and reproducing them. Social relationships and networks develop and change gradually (instead of jumping from 0 to 1 or from 1 to 0).

3. Why should social networks be central mediators of social processes, as a lot of research testifies?

Social networks are inert. Once in place, they transform slowly but channel communication effectively. For example, market actors tend to transact with the same suppliers or buyers, with the expectations from previous transactions reducing economic uncertainty.

4. Can we conceptualize and study networks between collective and corporate actors (like firms, states, or street gangs) in the same way as networks between individuals?

dentities feature as nodes in a network when they are identified as relevant actors in the communicative storytelling in the network. This mechanism does not differ between individuals and corporate or collective identities. All of these regularly serve as focal points of attribution and orientation in communication.

How does this approach differ from the other recent conceptualizations of the interplay of networks and social processes?

- Andrew *Abbott* advances a "processual sociology" with all social structures as "encoded" "memories" of past events (2001, 2016). However, he remains vague about how to conceptualize these processes (and how they leave traces of memory) and does not deal with social relationships and networks in particular.
- Harrison *White* argues since the mid-1990s that "switchings" between different network domains make for fresh connections and meaning (1995a; Mische and White 1998; Godart and White 2010; see section 7.8). It seems prudent to regard switchings as a special case of a communicative event that combines different sociocultural contexts, e.g., when communication involves multiple network ties. But White sees all social process as switchings (Schmitt and Fuhse 2015, 182f), making it hard to pinpoint the meaning of the concept. Also, if switchings incessantly make for new meaning, how can we ever observe the relatively inert social constellations (network domains, disciplines, institutions, control regimes) at the heart of White's theory?

- Mustafa *Emirbayer* in his "Manifesto" (1997) and Charles *Tilly* (2002, 2005, 2008) write that durable social patterns form in the course of "transactions" (see section 7.7). Again, this concept remains underdeveloped—what do transactions entail, and how do they make for the development, stabilization, and change of social patterns?

- Picking up on Luhmann, Stephan *Fuchs* writes of networks as a form of self-organization of communication (2001a, 251ff). Like me, he regards networks as patterns of meaning that stabilize and make for the definition of nodes and the relations between them. He goes on to elaborate on networks of culture (2001a, 280). For example, theories are networks that relate concepts in particular way, with a concept (say, "network") acquiring its meaning through its relations to other concepts. Social networks (of relationships between actors) and cultural networks (between symbols) are treated as similar here, without specifying their interrelation: For example, how do theory groups develop theories, and how are they held together by them? Also, the processes at play are not broken down to events of communication. These would have to forge, or to dissolve connections between actors and between symbols.[6]

- Mirroring Luhmann's "autopoietic turn," *John Padgett* picks up on models from chemistry to model the emergence and development of durable social structures such as organizations and networks (2012a, 2012b; Padgett and Powell 2012). He regards them as a "form of life" stabilizing in dynamic feedback processes in communication, similar to chemical "autocatalysis" (2012b). Padgett offers a formal model of "relational ties" as deriving out of "transactional success" in the "product flows" between individuals or organizations (2012a, 94a). This is more similar to network exchange theories than to communication theory as considered here. The link to communication comes in when Padgett considers symbols as part of this product flow, and with individuals and organizations as "symbol-transforming machines" (2012a, 96ff). Language, in a very limited sense, develops out of the flow of symbols between multiple actors.

  All of this then culminates in "network folding mechanisms" such as "transposition," "conflict displacement," and "multivocality" (Padgett and Powell 2012, 12ff). Padgett observes these mechanisms as linking

---

[6] I agree that social and cultural networks bear similarities in formal structure and in substance. However, I cannot offer a theory of cultural networks or detail their interweaving with social networks. For some preliminary considerations see Fuhse et al. (2020).

different networks and making for repercussions between them in his empirical studies. For example, a mismatch between familial and economic networks led to political turbulence in the rise of the Medici in Florence (Padgett and Ansell 1993). With the advancements from 2012, Padgett offers a full-fledged theory of organizational change out of multiple networks. However, he builds on a highly reduced notion of communication as the flow of products or symbols. My approach goes further in considering communication as the socialprocessing meaning that results in different kinds of social and cultural structures, and it is less abstract and formal than Padgett's chemistry-inspired models.

Overall, these various advances point in a similar direction: Social relationships and networks form, stabilize, and change over the course of communicative events. They constitute one form of "memory" (Abbott, Schmitt) through which events in the past have an effect on present communication. But most of these advances do not really elaborate on this memory, or on what aspects of communication are relevant for relationships and networks. In my approach, past events leave multiple traces that current events draw on, including individual identities, cultural institutions, and factual knowledge, as well as social relationships and networks. These result from the attribution of communicative events to actors, and from observing them as "messages" that indicate how actors (want to) relate to each other.

Of course, communication theory does not capture all there is to social networks and their dynamics. Any theoretical account only offers a particular perspective that allows us to see certain things (see section 1.3). In this case, the focus is on the production and modification of relational expectations in the sequence of communicative events. Other things— like the impact of individual cultural imprints on network processes, or the advantages or disadvantages of particular network positions—would better be modeled with action theory. No perspective gives us a full view of any phenomenon.

I have briefly discussed methods connected to this theoretical perspective in an excursus (section 8.10). Instead of readily reducing social constellations to stable patterns of 1s (ties) and 0s (non-ties), we should trace their occurrence in communicative events and their development in communicative sequences. This leads to the combination of network research with approaches studying naturally occurring communication, such as conversation analysis or interactional sociolinguistics. These give us insights into the relational meaning of linguistic forms and types of communicative events: How do they relate actors in a participation framework? These can then be typified and quantified for formal study, be it in relational event models or in the identification of network ties from event regularities. This approach is well suited for process-generated data on communication, as frequently studied in computational social science and digital humanities. Major methodological challenges concern the examination of multiple kinds of events and their potential cross-effects, as well as the disentangling of short-term dynamics of interaction and relatively stable relationships.

9

Summary and Discussion

Time to wrap things up. In this concluding chapter, I first summarize the main arguments, from the interplay of networks and meaning (section 9.1) in the process of communication (section 9.2) to specific forms of meaning in their interweaving with network patterns: relationships and identities (section 9.3), relational institutions (relationship frames, institutionalized roles, social categories; section 9.4), and group identities (section 9.5). Section 9.6 gives a brief overview of the premises and particular contributions of the theoretical perspective. In section 9.7, I address two lines of social theoretical objections. And section 9.8 discusses three areas for further development of the perspective: formal organization, fields of society, and the new data sources and methods of computational social science.

9.1. Networks and meaning

This book offers a theory of social networks, as interwoven with meaning, and as constructed in the process of communication and structuring it. This leads to the dual model of networks laid out in section 2.1:

(1) On the one hand, networks feature discernible *patterns of communication*: who communicates how with whom, and how frequently? Relationships are characterized by typical mixtures of activities—and this is how we distinguish ties from non-ties, but also different kinds of relationships (types of tie).

(2) On the other hand, these patterns of communication are structured by *meaning structures*. These consist of *relational expectations* about how particular actors will, and ought to, behave toward particular others. This includes the definition of relationships, and of the identities of actors in relation to each other.

Social Networks of Meaning and Communication. Jan Fuhse, Oxford University Press. © Oxford University Press 2022.
DOI: 10.1093/oso/9780190275433.003.0009

The two sides of social networks are intricately interwoven. Relational expectations develop in the process of communication from its observation with regard to underlying dispositions (see chapter 8). The expectations in a relationship are both cognitive and normative, in Johan Galtung's terms (1959; see section 7.3). It is expected that Carl *will* invite his good friend Marcus to his birthday party, but also that he *should* do so. A non-invitation will be seen as violating the norms of friendship and change the definition of the relationship between them. Therefore, relational expectations do not only map communication. They also structure them, making communication between alter and ego more predictable and reliable, and reducing the inherent uncertainty of all communication (see section 2.4).

While communication patterns and relational expectations are strongly linked, they constitute distinct aspects of social networks. On a conceptual level, communication is a process, whereas expectations constitute relatively durable structures underlying it. But the two should also be kept separate for substantial reasons: Communication patterns and meaning structures are affected by different sociocultural factors, and this makes for their relative independence. Opportunities for contact in schools, at the workplace, in the neighborhood, and other foci of activity have an impact on patterns of communication, simply by bringing people repeatedly together. This can lead to the formation of social relationships, and potentially to wider meaning structures—for example, when inner-city youth form rivaling street gangs by neighborhood. Conversely, institutionalized cultural patterns like social categories (ethnicity, gender) and blueprints for relationships offer templates for the organization of social relationships. They become incorporated into the meaning structure of social networks, thus symbolically ordering patterns of communication.

9.2. Communication

In principle, these arguments can be based on a number of concepts for the events in networks. The notions of action, practices, interaction, and transactions all consider the processing of meaning (in contrast to the concepts of behavior and exchange). We could use all of them to construct theories of how the meaning structures of relationships and networks develop, stabilize, and change over the course of events. However, I argue in chapters 7 and 8 that the concept of communication fits this task particularly well. It avoids inferring the subjective meaning of actors, instead focusing on the *processing of meaning in communicative events*. This leads to the methods sketched in section 8.10. Rather than ask people about their relationships to others, we can reconstruct them from the communication between them.

Following Niklas Luhmann, all communication is routinely monitored for factual information and for the message sent from a speaker or writer to the actor(s) addressed. This *message* aspect is most relevant for the construction of relationships and networks: Communicative events are routinely attributed to actors, leading to questions about their dispositions behind these events. This leads to expectations about these actors (identities) and about their relationships to others (relational expectations; sections 8.4–8.6). A phone call, a greeting, an invitation for a birthday party, a critical remark at a conference, and a diplomatic note will all be monitored for relational messages: How do alter and ego stand toward each other? Is the behavior by ego in line with the previous "definition of the relationship," or does her behavior show a disposition to change that definition? From these observations, expectations about the behavior by ego toward alter develop, as well as expectations about the expectations that ego holds of alter. Is alter now expected to reciprocally invite ego to her birthday party? Should she counter the critical conference remark? All events involving alter and ego are observed for whether they fit the expectations at place or deviate from them—making for their confirmation and reproduction, or for their change. As a consequence, social relationships and networks are relatively durable structures—a "memory" of communicative events—that guide communication, while at the same continuously changing.

9.3. Relationships and identities

Relationships and identities are the main ingredients of the meaning structure. *Relationships* are built on a relational definition of the situation, determining how alter and ego stand toward each other and what kind of communication is appropriate between them (see section 2.4). These relational expectations are complementary, but not necessarily reciprocal. Relationships can well be asymmetric, prescribing different activities and assigning roles to alter and ego.

The *identities* of actors constitute the second aspect of the meaning structure of social networks (see section 2.6). They are intricately linked to relationships and relationship frames: Alter and ego become "lovers," "friends," or "patron" and "client" by adopting the respective relationship frames. This attaches particular expectations to them, in line with those in the relationship. In sociology, such expectations are frequently captured with the concept of *roles* (to be discussed further in this chapter). The notion of identity is confined to singular actors and includes the idiosyncratic expectations negotiated in their relationships, as well as the multiple roles they occupy in relation to others. An actor can be loving husband, promising graduate student, mediocre player in the tennis club, singer in a punk rock band, friendly neighbor, and amusing storyteller among friends.

The construction of such multifaceted identities makes for a certain attuning of communication across relationships and social contexts (see section 2.6). And it couples subjective and communicative meaning: On the one hand, actors see themselves as embedded in these social contexts as part of their identity, and as affected by the expectations aimed at them. On the other hand, the irritations from individuals are observed in the social world in the construction of identities. People are seen as witty or boring, as committedly involved in one context, or as distracted by other concerns and obligations. All of this enters the meaning structure of social networks, assigning a social place to actors and affecting their relationships to others.

The meaning of relationships and the expectations attached to identities are not confined to dyads: If the communication between Betty and Lisa defines their relationship as "love," this has repercussions for their relationships to others. Their families and friendship circles have to accommodate this new relationship in their communication and in expectations. Similarly, if Marcus becomes the leader of his street gang, his network position changes, with the ties of other gang members to him and those between fellow gang members adapting to the change in his identity. Communication also frequently takes place with more than two actors, and thus with multiple ties, involved (see section 2.5). The construction of identities and of relationships incorporates this manifold coupling of processes across dyads, leading to meaning structures spanning *networks*, rather than only singular relationships.

9.4. Relational institutions

The negotiation and definition of identities and relationships builds on widely available cultural models. These include relationship frames and institutionalized roles, but also social categories. All of these provide cultural packages of relational expectations between actors—they are "relational institutions" prescribing how types of actors typically relate.

(a) *Relationship frames* such as "love," "friendship," "family," "patronage," and "competition" are blueprints for the organization of social ties (see section 2.7). They prescribe typical combinations of kinds of communication that are appropriate and expected in a given kind of relationship ("action profiles" in the sense of John Levi Martin). These relationship frames give kinds of interaction a relational meaning: a birthday phone call, an invitation for dinner, an attack in political debate, a colleague calling me "friend"—all of these incidents are connected to institutionalized packages for how to relate. Following the "relational work" approach by Viviana Zelizer, relationships are carefully maintained and negotiated in signals. And frequently certain kinds of interaction are avoided because they do not belong into the established kinds of relationships. The communication in relationships frequently adopts such relationship frames, but with considerable wiggle room for modification and combination.

(b) Relationship frames are frequently coupled to institutionalized *role frames*, such as "friend," "husband," "patron," or "student" (see section 2.7.d). By virtue of forming a friendship, a marriage, or a patronage tie, fixed bundles of expectations are attached to an actor. As argued in section 5.2, roles are inherently relational. They consist of what is expected of an actor in relation to other actors or, rather, of the expectations attached to a position in relation to other positions. These role expectations pertain to multiple social ties at once: My friend is supposed to treat not only me, but also my friends amicably. A husband has to interact in particular ways with his wife, but also (in other particular ways) with her friends and family. A teacher has to treat her students as fits her role, but also their parents and her colleagues. Relationship frames and roles are thus not confined to the dyadic level but make for a symbolic ordering of network patterns into patterns of "structural equivalence" (following Harrison White).

Roles are a tricky concept to add to the theory. On the one hand, they capture systematic relationship patterns by recognizable positions in social networks. On the other hand, they stand for cultural blueprints for these positions and relationships between them. Therefore, I term these cultural blueprints "roles frames," and the actual role patterns in social networks can come either from such cultural prescriptions or emerge endogenously as recognizable patterns of communication. Another confusing problem stems from the frequent entanglement of role frames in formal organization. Roles like professor and student or clerk and customer form part of the administrative world of universities, companies, and states. While they come with culturally set bundles of relational expectations, they are not strictly negotiated on the dyadic level. Rather, these positions are assigned, and the precise content of expectations attached to them determined, in organizational decisions, often far removed from the actors in these positions. This opens up a Pandora's box of conceptual questions that I leave untouched in this book (see section 9.8.a).

(c) *Social categories* like ethnicity and gender constitute a third kind of relational institution. Similar to institutionalized roles, they come with cultural prescriptions for the kinds of interaction within and across the category. Ethnic categories divide people by (supposed) common descent and frequently foster positive personal relationships between co-ethnics, whereas cross-category personal ties are discouraged or shunned (see chapter 4). Ethnic categories do not only make for structural separation—they also regulate the interaction across the category. However, the salience of an ethnic category depends on communication patterns following its lines. The more ethnic groups live in different neighborhoods, attend different schools, and occupy different occupations, the more convincing it is to classify into "different kinds of people (that better not mingle)." Nowadays, the symbolic division of people by ethnicity has lost legitimacy in cultural discourse. But ethnic categories should only disappear if foci of activity and network patterns become less segregated.

The *gender* category has also come under pressure but continues to regulate interaction and relationship formation (see chapter 6). Unlike many ethnic categories, it does not make for a structural separation of positive personal relationships. Rather, it tends to prescribe different kinds of relationships between members of the same gender and those of different genders: Friendships primarily run among men or among women. Love relationships, in contrast, are still mostly formed between women and men. The genders thus occupy positions of "structural equivalence" (or structural similarity) in the network of personal relationships. Again, we find a coupling of (gender) roles to relationship frames, in this case to the frame of heterosexual love that partly defines the genders in relation to each other.

Overall, these relational institutions form part of *cultural repertoires* (see sections 2.1 and 2.2). These repertoires are connected to the communication in networks and to a certain extent depend on their adoption there. Symbols have to be disseminated but also put to use in networks; otherwise, they fail to convince as meaningful. However, cultural repertoires become relatively independent of networks through their diffusion in mass media, economic markets, state bureaucracies, and systems of education. Networks are interwoven with culture, but both do not collapse into one due to this relative independence (in line with Emirbayer and Goodwin 1994).

This independence also implies that social relationships and networks have a certain leeway in adopting these relational institutions. First, the communication in a relationship can decide which cultural models to follow. It can define the relationship as love, friendship, or rivalry; it can assign different institutionalized roles to actors; and it can follow the expectations tied to social categories or ignore them. Second, different institutions can be modified or combined in the negotiation of relational expectations. For example, two people can become "friends with benefits" or have an "affair," rather than define their relationship as either "friendship" or "romantic love" with the full package of expectations tied to these cultural frames.

In principle, communication in a social relationship is autonomous to develop its own set of relational expectations. However, a relationship has to fit into the wider context of other social relationships and obligations. And it is interactionally less demanding to simply follow preestablished cultural expectations. According to Talcott Parsons, a shared culture effectively reduces the inherent uncertainty in communication. It helps us to roughly know what to expect of each other, and it helps communication to proceed without grave misunderstandings and conflicts. It is easier to submit to the "recipe knowledge" (Berger and Luckmann) of social relationships than to deviate from it, or to work out relational expectations from scratch. To summarize, cultural institutions constitute a relatively *autonomous* level of the social world. But they are also *indeterminate* in the sense of not being able to fully structure what is going on in networks.

Finally, like all cultural patterns, relational institutions *vary* across social-cultural contexts. As King-To Yeung shows, the concept of "love" differs in its meaning between urban communes (2005). Social categories like ethnicity and gender have varying expectations attached (and acquire a different salience and force) by countries, social milieus, generations, contexts, and historical times. Similarly, patronage was a powerful model for social relationships in Renaissance Florence and may still be in force in social contexts like Southern Italy and academia. But it is less powerful and convincing in labor markets (supposedly) structured by meritocratic achievements. Also, the relationship frames of romantic love and of friendship are distinctly modern constructions. They seem to develop once the rigid social structures of feudalism make way for the more fleeting and less reliable social relations of modern society.

9.5. Group identities

Group phenomena are a curious instance of semi-institutionalized sociocultural patterns (see chapter 3). Group cultures develop out of close-knit local meshes of social relationships, often formed around common activities at foci. In their repeated contact, they arrive at shared symbols that distinguish them from other groups and from the outside world. And they acquire a sense of commonality, of collective identity, often defined against others (out-groups). This collective identity symbolically draws a sharp distinction between inside and outside and comes with expectations for how to treat fellow group members and outsiders. In this regard, group identities and boundaries are similar to social categories, but they are much more confined to a relatively small number of actors that know each other face to face.

Structurally, the group is differentiated gradually into a core of devoted followers, including the group leaders, on the one hand, and a periphery of individuals with mixed commitments and with crosscutting cultural influences on the other hand. Here, the distinction between actual group members and interested bystanders blurs. A group frequently demands full loyalty by its members, but it can never clearly tell those belonging to the group from others (unlike formal organizations). The sharp distinction between inside and outside remains symbolic and does not correspond to a clear structural separation.

Many groups draw on institutionalized models to define themselves. For example, local ethnic-minority youth in deprived inner-city neighborhoods often define themselves as "street gangs," building on models of groups around them and disseminated by the mass media. These models are relational institutions that structure social relationships within the group, but also its relations to the outside world, including other groups (e.g., rivaling street gangs). A multilevel architecture of networks results, with networks of social relationships inside the group, but also between groups as higher-level actors connected in ties of alliance or conflict.

However, the installment of a group identity and a symbolic boundary to the outside itself constitutes a gradual institutionalization. At the lower end of the spectrum, communication is primarily oriented toward its individual members, building on and defining the relational expectations between them. Moving upward toward institutionalization, communication is increasingly framed in terms of the group identity, and individual members become less and less important. In a fully established street gang, a leader can be imprisoned or killed and replaced without putting the group survival in jeopardy. Group identities and boundaries are thus gradually accomplished and make for a group's "life of its own." The notion of institutionalization is here confined to the group context, where cultural models for how to behave to fellow members and to nonmembers gradually replace (or complement) the relational expectations negotiated in idiosyncratic relationships. In order to become a full-fledged relational institution (as coined in chapter 5), the group identity has to be taken up by the mass media and adopted by other groups.

Groups in the sense of chapter 3 are the exception, rather than the rule, in contemporary social structure—counter to the "groupism" of German formal sociology, of symbolic interactionism with Gary Alan Fine as the currently most prominent advocate, and of the "relational realism" of Pierpaolo Donati and Margaret Archer. Social structure can be better be termed as networks of interrelated social relationships. These rarely (and only gradually) crystallize to persistent group structures. The formation of groups depends on foci of common activity and on the gradual establishment of a group identity. But their members remain tied to other social contexts through crosscutting memberships. Even violent and radically deviant groups such as street gangs are not able to cut their members completely out of their wider webs of personal relationships.

9.6. Premises and contribution

Overall, this book adheres to three theoretical premises:

(1) The social world is filled with *structures of meaning*, including identities, social relationships, networks, groups, roles, social categories, and institutions, but also formal organizations and social fields like politics, science, and the economy.

(2) Diverse structures of meaning, including social relationships and networks, develop over the course of ephemeral *social events*, and they guide these events in turn.

(3) *Meaning* can fruitfully be conceptualized and studied as processed *in communication*, rather than as primarily (or only) held subjectively by individual actors.

The first two assumptions are variously shared by action theorists Max Weber, Alfred Schütz, and Talcott Parsons; symbolic interactionism; Niklas Luhmann's systems theory; and Pierre Bourdieu's theory of practices and fields. Like Weber, Parsons, and Luhmann, I use the concept of "expectations" to denote the relatively durable patterns of meaning. The third premise follows Luhmann's theory of communication and parts with the theories of action, symbolic interaction, and social practices. The assumption is that communicative events carry meaning in and by themselves, in line with what Robert Wuthnow calls the "dramaturgical approach" (1989). However, I do not deny that people do process meaning in their heads, and that communication breaks down without the individuals involved understanding linguistic forms and other signs processed in communication in roughly similar ways. In a sense, the decision to remove subjective meaning from the theoretical perspective springs from conceptual and methodological convenience, rather than from ontological conviction.

Based on these premises, I conceptualize social relationships and networks as patterns of meaning that develop in communication and structure it. Of course, I am not the first, nor the only one advancing this perspective. In my partial view, the basic stance resembles that of a number of formulations in relational sociology: Harrison White's theory of social networks, Mustafa Emirbayer's "Manifesto for a Relational Sociology" (not his writing on agency), Charles Tilly's arguments since 1998 about basic social forms deriving from transactions, Stephan Fuchs's theory of cultural networks, and John Padgett's theory of network-folding mechanisms. Andrew Abbott adopts a similar perspective, but without the focus on networks. More than White, Tilly, and Padgett, I flesh out a theoretical architecture firmly built on established sociological concepts—allowing for better accessibility and for the possibility of scrutinizing my efforts within theory discourse. Unlike Fuchs (and actor–network theory [ANT]), I detail the particular mechanisms in *social* networks, rather than subsume all kinds of patterns of relations under a broad and vague rubric of "networks" (see section 9.7). In contrast to Emirbayer's "Manifesto," I regard social relationships and networks not as mere descriptions of processes (see section 7.7), but as relatively durable structures of expectations deriving from the process and guiding it.

What are the particular contributions offered?

- I advance an *integrated framework* for in the interplay of networks, meaning, and communication (chapter 2). This separates communication patterns, the meaning structure of networks, wider cultural patterns, and opportunities for contact to address their interplay.
- The meaning of social relationships and networks is specified as *relational expectations* (sections 2.4 and 8.6).
- Instead of relying on seemingly essential differences between different types of tie ("love," "friendship," "patronage," etc.), I argue that these stem from specific cultural models for kinds of relationships: *relationship frames* (section 2.7 and Fuhse 2013). This also explains their varying structural propensities.
- *Groups* are conceptualized as an ideal type of *proto-institutionalized structures of meaning* at the meso-level, rather than as basic building blocks of the social world (chapter 3).
- The interplay of *institutions* and network patterns is fleshed out (chapter 5). Social categories, relationship frames, and institutionalized roles are examples of "relational institutions" that bear an imprint on networks of social relationships.
- I offer a theory of the connection of the *gender category* to networks of personal relationships, in conjunction with the relationship frame of "heterosexual love" (chapter 6).
- The precise development of *relational expectations from communicative events* is traced to their attribution to actors as "messages" sent to other actors (chapter 8).

Overall, then, I see these as reasons enough to write this lengthy book. Of course, the reader will not necessarily agree with all of my assessments or my conceptual choices. But hopefully, she or he will find something worth engaging with (critically), or using for her or his own research. After all, this is what theory is for: to give us a perspective from which to make sense of empirical observations, and to ideally guide them.

9.7. Objections

These arguments do probably not sway everybody. The basic construction of social networks between actors formed and developing over the course of communication faces criticism from two sides. On the one hand, a diverse phalanx of critical realists and of pragmatism-inspired authors insists on a central place for *human beings* and *agency* in social theory. From a critical realist perspective, Christian Smith faults relational sociologists like Harrison White for a "reductionist" view of social reality and for not providing an answer to the question "what is actually real about human being" (2010, 11). Pierpaolo Donati and Margaret Archer extend the critical realist ontology to include social relations as "realities sui-generis" (2015, 8ff, 19ff; see section 3.6). In a litany against other relational sociologists, they treat my approach relatively mildly:

> Fuhse has come to treat social relations [. . .] as mere communications. (Donati and Archer 2015, 22; see also Ruggieri 2014)

From a pragmatist standpoint, Mustafa Emirbayer and Jeff Goodwin criticize structuralism in general, and Harrison White in particular, for not taking "human agency" into account (1994, 1437f, 1442ff; see also Emirbayer and Mische 1998; Burkitt 2016). A number of relational sociologists even in the circle around White take agency to be of central importance in the social world (McLean 2007, 6, 118f, 121; Erikson 2013, 233ff).

On the other hand, ANT and François Dépelteau relegate human beings to a similar position as animals, material objects, technical artifacts, and sometimes also concepts and ideas—as actors related to other actors and thereby constituted. In this vein, Bruno Latour formulates his principle of symmetry: human beings should not be conceptualized and studied differently from the things around them (2005, 89). The relations become primary or, rather, the processes of relating. Building on James Dewey and Arthur Bentley's concept of "transactions" (see section 7.7), Dépelteau argues that we should focus our theories and our research on "complex transactions or (or associations) between various human and non-human transactors" (2015, 49). Stephan Fuchs similarly adopts a wide view of the entities related in networks (2001a, 251ff). While networks can incorporate all sorts of things as nodes, his main concern lies on cultural networks of relations between symbols or ideas.

The two lines of attack clearly stand in conflict. Critical realists and pragmatists want to reserve a special place for human beings in sociological theory. ANT, Dépelteau, and Fuchs, in contrast, see networks (or relating processes) between all kinds of entities—and these entities, including human beings, become secondary to the processes between them. My own position and similar approaches (White, Tilly, Padgett) are caught between these two extremes.

To counter the first line of attack, I formulate a first constructivist credo: our theories are not "accurate representations" of the social world and do not have firm ontological certainties to build on. Of course, we can proclaim individuals and social relations (whatever this includes) as real, as Donati and Archer do. But this supposes that we somehow "know" out of philosophical consideration the world around us, and that we undoubtedly have the right concepts at hand (Cruickshank 2010). In contrast, I see the construction of sociological theories as a risky enterprise, prone to run astray or into dead ends. This includes the possibility that my own theoretical approach gets stranded. But this has to be decided by scrutinizing its consistency and parsimony, and by confronting it with empirical research, rather than by ontological proclamations.

The pragmatist position is more nuanced than critical realism, and it comes in different varieties (cf. Rorty [1979] 2009; Putnam 1981). Generally, pragmatism holds that social science is part of the interaction in human communities. This argument can lead to an insistence that we should take human beings seriously and put them center stage in our reflections on social life, or that we follow lay conceptions of the social world in our reconstructions. Richard Rorty even wants to give up on abstract theorizing ([1979] 2009; Reed 2011). Unlike critical realism, these reflections do not amount to an ultimate claim for truth, only for scientific activity in closer connection to the social community. Rightly so, because overcoming or correcting lay understandings lies at the heart of scientific activity. According to Jeffrey Alexander, it is wrong to maximally contrast social life and abstract theorizing—the construction of theory is as much part of communal life as lay language and understandings (1995, 119). Looking at Dewey and Bentley's essay "Knowing and the Known," I fail to see how the pragmatists adhere to the precepts of humanism and of sticking to the language of their wider community: The notion of "trans-action" introduces new vocabulary at odds with lay understandings. And Dewey and Bentley do not put human beings and their agency center stage. This very much adheres to Rorty's "project of finding, new, better, more interesting, more fruitful ways of speaking" ([1979] 2009, 360). As does my own attempt to ground sociocultural constellations in the concept of communication.

My defense against the challenges of ANT, François Dépelteau, and Stephan Fuchs is again constructivist, but of a different kind than theirs. ANT deconstructs the notions of "actor" and "social relationship" by claiming that actors (or, rather, actants) are not actually driving social processes, but constituted by them, and that we have to disaggregate relationships into such processes. So far I agree, and relational sociology says much the same (Mützel 2009). However, ANT mostly stops with this *de*construction instead of continuing with the *re*construction of what we mean by actors and relationships.

In my approach, actors become meaningful projection points in the social world and the subject of (relational) expectations by the attribution of communicative events to them. This happens in storytelling, as elaborated by Tilly (2002, 2006), but also implicitly in the observation of communication with regard to its message aspect (see sections 8.4–8.6). Social relationships result from this process as bundles of relational expectations that effectively guide future communication. As a consequence, social relationships are relatively durable meaning structures, rather than only processes. These run between the kinds of things we attribute action to: individuals, street gangs, companies, social movements, universities, and states (see section 8.9).

These considerations make for the qualitative difference between social networks and other kinds of networks, be they sociomaterial networks between objects and individuals, or cultural networks between concepts or other symbols. I do not here engage with the question of whether these other kinds of networks show similar propensities for inertia and for effects in the social world as social networks. But I do argue against simply conflating different kinds of relations between heterogeneous entities in metaphors of "networks" and "associations" that have to remain as vague as they are broad.

However, none of these arguments suffices to straight out dismiss alternative theoretical approaches. As argued in section 1.3, I regard theories as perspectives that allow us to see and to observe some aspects of a complex social reality (Giere 2006a). No theory can be rejected on the basis of arguments drawn from a different theoretical framework. Rather, it should be judged on the basis of its internal consistency, its parsimony, and its ability to make sense of empirical research and to guide it. I will not engage in the discussion of other theories in these regards here but only clearly state the criteria that I want my theoretical perspective to meet and to be measured against.

9.8. Going forward

To conclude this book, I briefly discuss three areas for the theoretical perspective to extend to. The first two of these cover social phenomena that I have at times touched on but not engaged with systematically in this book: organizations (a) and fields of society (b). The third addresses the new kinds of data and research questions opened up by the developments in computational social science (c).

(a) Formal organization

First, I do not yet have a satisfactory answer to the question: How are formal networks of social relationships and *formal organization* related? Obviously, organizations like companies, universities, state bureaucracies, universities, schools, but also sports clubs and voluntary associations are foci of activity that make for repeated contact and foster social relationships among their members. Conversely, this "informal organization" has an impact on the formal organization: on hiring and promotions in companies and universities, on mobilization and identification in social movement organizations, on organizational performance, and on conflicts within (Burt 1992, 2004; Ibarra 1992; Gould 1995; Krackhardt 1999; Lazega and Pattison 1999; Diani and McAdam 2003; Baldassari and Diani 2007). However, we still lack a clear conceptualization on the relation of networks and organizations.

One option is to treat the two as separate realms of the social world. This follows my considering social relationships as patterns of relational expectations that develop in the course of communication between two actors (sections 2.4 and 8.6). We speak of a social relationship only if, and to the extent that, patterns of communication and relational expectations differ from formally prescribed role relations. Communication between waiter and client, or between professor and student, can follow the cultural scripts attached to these roles, or they can differ from them on the basis of a social relationship. In this conceptual option, social relationships and formal organization would be two different structures of expectations that simultaneously affect communication—and thereby each other (see section 7.1).

Alternatively, we can treat formal organization as an institutionalized pattern that becomes incorporated in networks of social relationships and thereby structures it. Social network would then be a master concept for the relating of actors, and formal organization would be incorporated in this conceptual frame like friendship, kinship, romantic love, and patronage. This follows neo-institutionalism (Meyer and Rowan 1977) and is consistent with my arguments about institutions and networks in chapter 5. Both options seem feasible to me but come with wildly diverging conceptual implications. With the first option, we could speak of networks *and* organizations; in the second option, we would have formal organization *in* networks. Both options would have to be spelled out in detail and confronted with empirical research to decide between them.

(b) Fields of society

A second area of extension concerns social fields at the macro-level, like politics, the economy, science, law, art, etc. How should we conceptualize the role of social networks in these fields and, conversely, the impact of these fields on networks? Niklas Luhmann lays out these spheres of society as "functional subsystems" ([1997] 2013, 65ff). Unlike Parsons, he does not devise them as fulfilling objective functions for society, but as self-reproducing spheres of communication with their own codes, media, and criteria for applying them: payments and prices, power and political programs, scientific truth and theory/methods, and so on. In principle, the identities of actors and the social relationships between them do not matter for the processing of these codes. Beggar or prince, all have to pay the same prices on economic markets and are subject to state power and to the same impartial laws, and their artistic creativity and scientific ideas are judged by the same impersonal criteria. Generalized media like money, political power, law, and scientific truth make communication work without social relationships. In a sense, they replace them as facilitators of social coordination. Luhmann consequently does not consider social relationships and networks in his functional subsystems.

Developed at the same time, Pierre Bourdieu's theory of fields makes for an antithesis to systems theory (Bourdieu and Wacquant 1992, 71ff). He views fields like the economy, the state, or science as subject to struggles between actors with regard to the distribution of resources (capitals) and to the symbolic criteria governing these struggles. Here the relations between actors become a central focus, both in terms of their relative positions in fields and of their symbolic practices marking social vicinity or distance. Actual social relationships and networks could play an important role here, but they remain marginal in Bourdieu's theory (Bottero 2009; Mohr 2013; see section 7.4).

Recently, a number of authors have picked up on field theory to incorporate a more systematic concern for networks (DiMaggio 1986; Anheier, Gerhards, and Romo 1995; de Nooy 2003; White et al. 2004; Powell et al. 2005; Beckert 2010; Bottero and Crossley 2011). Much of this is wedded to the neo-institutionalist perspective developed around Paul DiMaggio (see section 5.6). Networks between field actors are seen as governed by institutional rules emerging "isomorphically" out of the mutual observation and orientation in a field. The range of phenomena covered by this concept of fields remains wide and vague, including organizations, local arts scenes, economic markets, scientific disciplines, etc. Roger Friedland and Robert Alford lay out the "institutional logics" perspective firmly focused on macro-spheres of society (1991). Their list of "institutional orders" looks surprisingly similar to Luhmann's functional subsystems: capitalism (economy), democracy and the state (politics), religion, science, and the family. Each of them has its own central logic—a "set of material practices and symbolic constructions [. . .] available to organizations and individuals to elaborate" (Friedland and Alford 1991, 248).

These approaches suggest that social macro-spheres affect social relationships and networks in the form of institutions. For example, the rules of competition and conflict in economic markets, in multiparty democracy, in the scientific field, and in the art world render the formation of particular relationships and network constellations between actors (companies, political parties, politicians, scientists, universities) likely (Fuhse 2020a).

The "family" constitutes a special case of a sphere strongly structured by cultural prescriptions but segmented into small-scale units of familial relations (see section 6.5). In all instances, we can observe a symbolic ordering of networks by their connection to one or another sphere of society: political alliances follow different logics than scientific collaborations or economic competition. Social fields seem to make for different "types of tie" with their own structural tendencies and with different network positions bringing advantages in the field.

(c) Computational social science

The third direction forward does not really involve social phenomena not yet addressed. Rather, it results from the development of new methods and of new data sources in what is called "computational social science" and "big data" (Lazer et al. 2009). In principle, this methods-heavy area of research covers all sorts of phenomena, and it can be used to address questions from quite different perspectives. For example, we could examine online sales or online dating focusing on individual traits, leading to conventional statistical analyses. Or we can apply formal network analysis to discern structural patterns between consumption items or in dating interactions. However, I see the developments in computational social science as particularly relevant for the perspective in this book because they allow us to address its two key issues: the interweaving of social networks with meaning and the development of social relationships and networks in communication.

The relevance of computational social science is more obvious for the second issue: Whether in online communication or in archival documents like scientific texts and parliamentary proceedings, the new data sources frequently include process-generated information on *time-stamped communicative events* successively reacting to previous events. These allow us to reconstruct structural patterns of relationships, but also the micro-dynamics of interaction in these events (Kitts 2014, 281ff; de Nooy 2015; see section 8.10). But of course, theory and methods have to be adapted to the data at hand. In the context of computational social science, we are frequently dealing with large number of events that render the qualitative interpretation and typification of the material impossible. There are two potentially complementary ways out:

- We have to rely on automated methods to investigate the meaning in these events, for example, by looking for keywords that signal support ("yes," "agree") or disagreement ("no," "but"). Such procedures might err in identifying the meaning of some events. But careful validation should tell us whether we are right often enough, or whether these methods are prone to systematic distortions.
- We can zoom in on some of these events with automatically detectable features to qualitatively interpret their meanings, in what Ronald Breiger, John Mohr, and Robin Wagner-Pacifici call "computational hermeneutics" (Mohr, Wagner-Pacifici, and Breiger 2015; Breiger, Wagner-Pacifici, and Mohr 2018). This should help us consider the meaning of quantitative results and avoid returning to pure structuralism.

Regarding the first issue, computational social science constitutes a push for quantitative methods and seems to run counter to an interest in patterns of *meaning*. This is true if one takes meaning to be only subjective: Large-scale quantitative analysis of process-generated data indeed takes a step away from the individual and her subjective understandings. Very rarely are individuals available for qualitative interviews to complement structural analyses. However, this book advocates a perspective on meaning as processed in communication. Here computational social science gives us rich opportunities for empirical study, for example, by way of quantitative text analysis (Bail 2014; Evans and Aceves 2016; Fuhse et al. 2020). This offers new ways of examining meaning in its systematic structures (Mohr 1998). Methods like topic modeling and formal network analysis of co-occurring terms have unprecedented firepower to discern cultural patterns (DiMaggio, Ng, and Blei 2013; Light 2014; Rule, Cointet, and Bearman 2015). A more refined look at meaning patterns is taken by studies that go beyond mere co-occurrence to detect sentence structure and prevalent ways of storytelling (Mohr, Wagner-Pacifici, and Breiger 2013; van Atteveldt et al. 2017).

If these methods discern patterns of meaning, we only have to relate these to networks of social relationships between actors. Examples for this are Christopher Bail's study on media discourse about Islam (2012) and the analysis of sociosemantic networks spearheaded by Camille Roth (Roth and Cointet 2010; Basov and Brennecke 2017). Social networks can be detected on the basis of available information like coauthorships in science or contacts on social networking sites like Facebook ("friends") or Twitter ("follower"). Or they can be reconstructed from sequences of communicative events (see section 8.10). Given the complex structures of process-generated data, these often allow us to analyze all three key aspects of the theoretical perspective in tandem: networks, meaning, and communication.

Daniel McFarland, Kevin Lewis, and Amir Goldberg suggest that computational social science can become a "trading zone" between different disciplines (2015, 24ff). Social scientists, computer scientists, linguists, and physicists all bring their different expertise into the encounter and benefit from that of others. Confronted with the superior modeling of physicists, the programming skills of computer scientists, and the linguists' knowledge about the formal structure of language—what do the social scientists contribute? Ideally, we should have a better substantial knowledge of the subject of sociocultural patterns, based on the theoretical debates and on the extensive empirical research in our disciplines.

References

Abbott, Andrew. 1995. "Things of Boundaries." *Social Research* 62 (4): 857–882.

Abbott, Andrew. 2001. *Time Matters.* Chicago: The University of Chicago Press.

Abbott, Andrew. 2016. *Processual Sociology.* Chicago: The University of Chicago Press.

Abend, Gabriel. 2008. "The Meaning of 'Theory.'" *Sociological Theory* 26 (2): 173–199.

Adams, Rebecca, and Graham Allan, eds. 1998. *Placing Friendship in Context.* Cambridge: Cambridge University Press.

Alba, Richard. 1990. *Ethnic Identity: The Transformation of White America.* New Haven: Yale University Press.

Alexander, Jeffrey. 1995. *Fin de Siècle Social Theory.* London: Verso.

Alexander, Jeffrey, and Bernhard Giesen. 1987. "From Reduction to Linkage: The Long View of the Micro-Macro Debate." In: *The Micro-Macro Link,* edited by Jeffrey Alexander, Bernhard Giesen, Richard Münch, and Neil Smelser, 1–42. Berkeley: University of California Press.

Anheier, Helmut, Jürgen Gerhards, and Frank Romo. 1995. "Forms of Capital and Social Structure in Cultural Fields: Examining Bourdieu's Social Topography." *American Journal of Sociology* 100 (4): 859–903.

Ansell, Christopher. 1997. "Symbolic Networks: The Realignment of the French Working Class, 1887–1894." *American Journal of Sociology* 103 (2): 359–390.

Archer, Margaret. [1988] 1996. *Culture and Agency.* Cambridge: Cambridge University Press.

Arora, Saurabh, and Bulat Sanditov. 2015. "Cultures of Caste and Rural Development in the Social Network of a South Indian Village." *SAGE Open.* doi: 10.1177/2158244015598813.

Arquilla, John, and David Ronfeldt. 2001. *Networks and Netwars.* Santa Monica: RAND.

Azarian, Reza. 2010. "Social Ties: Elements of a Substantive Conceptualization." *Acta Sociologica* 53 (4): 323–338.

Baecker, Dirk. 2005. *Form und Formen der Kommunikation.* Frankfurt am Main: Suhrkamp.

Baerveldt, Chris, Bonne Zijlstra, Muriel De Wolf, Ronan Van Rossem, and Marijtje Van Duijn. 2007. "Ethnic Boundaries in High School Students' Networks in Flanders and the Netherlands." *International Sociology* 22 (6): 701–720.

Bahrdt, Hans-Peter. 1961. "Zur Frage des Menschenbildes in der Soziologie." *Archives Européennes de Sociologie* 2 (1): 1–17.

Bail, Christopher. 2012. "The Fringe Effect: Civil Society Organizations and the Evolution of Media Discourse About Islam Since the September 11th Attacks." *American Sociological Review* 77 (6): 855–879.

Bail, Christopher. 2014. "The Cultural Environment: Measuring Culture with Big Data." *Theory & Society* 43 (3/4): 465–482.

Bailey, Stefanie, and Peter Marsden. 1999. "Interpretation and Interview Context: Examining the General Social Survey Name Generator Using Cognitive Methods." *Social Networks* 21 (3): 287–309.

Baldassari, Delia, and Mario Diani. 2007. "The Integrative Power of Civic Networks." *American Journal of Sociology* 113 (3): 735–780.

Bales, Robert. 1950. *Interaction Process Analysis: A Method for the Study of Small Groups.* Cambridge, Massachusetts: Addison Wesley.

Bandelj, Nina. 2012. "Relational Work and Economic Sociology." *Politics & Society* 40 (2): 175–201.

Barabási, Albert-László, and Eric Bonabeau. 2003. "Scale-Free Networks." *Scientific American* 288 (5): 60–69.

Barnes, J.A. 1954. "Class and Committees in a Norwegian Island Parish." *Human Relations* 7 (1): 39–58.

Barth, Fredrik. 1969. "Introduction." In: *Ethnic Groups and Boundaries*, edited by Fredrik Barth, 9–38. Oslo: Universitetsforlaget.

Basov, Nikita, and Julia Brennecke. 2017. "Duality Beyond Dyads: Multiplex Patterning of Social Ties and Cultural Meanings." *Research in the Sociology of Organizations* 53: 87–112.

Bateson, Gregory. 1972. *Steps to an Ecology of Mind.* Chicago: Chicago University Press.

Baxter, Leslie. 1987. "Symbols of Relationship Identity in Relationship Cultures." *Journal of Social and Personal Relationships* 4 (3): 261–280.

Bearman, Peter. 1997. "Generalized Exchange." *American Journal of Sociology* 102 (5): 1383–1415.

Bearman, Peter, James Moody, and Katherine Stovel. 2004. "Chains of Affection: The Structure of Adolescent Romantic and Sexual Networks." *American Journal of Sociology* 110 (1): 44–91.

Bearman, Peter, and Paolo Parigi. 2004. "Cloning Headless Frogs and Other Important Matters: Conversation Topics and Network Structure." *Social Forces* 83 (2): 535–557.

Beattie, Irenee, Karen Christopher, Dina Okamoto, and Sandra Way. 2005. "Momentary Pleasures: Social Encounters and Fleeting Relationships at a Singles Dance." In: *Together Alone: Personal Relationships in Public Places*, edited by Calvin Morrell, David Snow, and Cindy White, 46–65, Berkeley: University of California Press 2005.

Beck, Ulrich, and Elisabeth Beck-Gernsheim. [1990] 1995. *The Normal Chaos of Love.* Cambridge: Polity.

Becker, Howard S. 1963. *Outsiders: Studies in the Sociology of Deviance.* New York: Free Press.

Becker, Howard S. 1982. *Art Worlds.* Berkeley: University of California Press.

Becker, Howard L., and Ruth Hill Useem. 1942. "Sociological Analysis of the Dyad." *American Sociological Review* 7 (1): 13–26.

Beckert, Jens. 2010. "How Do Fields Change? The Interrelations of Institutions, Networks, and Cognition in the Dynamics of Markets." *Organization Studies* 31 (5): 605–627.

Bell, Robert, and Jonathan Healey. 1992. "Idiomatic Communication and Interpersonal Solidarity in Friends' Relational Cultures." *Human Communication Research* 18 (3): 307–335.

Bellotti, Elisa. 2015. *Qualitative Networks: Mixed Methods in Sociological Research.* Abingdon: Routledge.

Bentley, Carter. 1987. "Ethnicity and Practice." *Comparative Studies in Society and History* 29 (1): 24–55.

Berger, Charles. 1988. "Uncertainty and Information Exchange in Developing Relationships." In: *Handbook of Personal Relationships*, edited by Steven Duck, 239–255. Chichester: Wiley.

Berger, Peter, and Hansfried Kellner. 1964. "Marriage and the Construction of Reality." *Diogenes* 12 (46): 1–24.

Berger, Peter, and Thomas Luckmann. [1966] 1991. *The Social Construction of Reality*. London: Penguin.

Bergmann, Jörg. [1987] 1993. *Discreet Indiscretions: The Social Organization of Gossip*. New York: Aldine de Gruyter.

Billig, Michael. 1995. *Banal Nationalism*. London: Sage.

Blau, Peter. 1964. *Exchange and Power in Social Life*. New York: Wiley.

Blau, Peter. 1977. *Inequality and Heterogeneity*. New York: Free Press.

Blum, Alan, and Peter McHugh. 1971. "The Social Ascription of Motives." *American Sociological Review* 36 (1): 98–109.

Blumer, Herbert. 1969. *Symbolic Interactionism*. Berkeley: University of California Press.

Bochner, Arthur, Carolyn Ellis, and Lisa Tillmann-Healy. 2000. "Relationships as Stories: Accounts, Storied Lives, Evocative Narratives." In: *Communication and Personal Relationships*, edited by Kathryn Dindia and Steve Duck, 13–29. Chichester: Wiley.

Boissevain, Jeremy. 1968. "The Place of Non-Groups in the Social Sciences." *Man* 3 (4): 542–556.

Bommes, Michael, and Veronika Tacke. [2006] 2012. "General and Specific Characteristics of Networks." In: *Immigration and Social Systems*, edited by Christina Boswell and Gianni D'Amato, 177–199. Amsterdam: Amsterdam University Press.

Bonilla-Silva, Eduardo. 1997. "Rethinking Racism: Toward a Structural Interpretation." *American Sociological Review* 62 (3): 465–480.

Boorman, Scott, and Harrison White. 1976. "Social Structure from Multiple Networks. II. Role Structures." *American Journal of Sociology* 81 (6): 1384–1446.

Borgatti, Stephen, and Daniel Halgin. 2011. "On Network Theory." *Organization Science* 22 (5): 1168–1181.

Bott, Elizabeth. [1957] 1971. *Family and Social Network*. London: Tavistock.

Bottero, Wendy. 2009. "Relationality and Social Interaction." *British Journal of Sociology* 60 (2): 399–420.

Bottero, Wendy, and Nick Crossley. 2011. "Worlds, Fields and Networks: Becker, Bourdieu and the Structures of Social Relations." *Cultural Sociology* 5 (1): 99–119.

Bourdieu, Pierre. [1980] 1990. *The Logic of Practice*. Cambridge: Polity.

Bourdieu, Pierre. 1986. "The Forms of Capital." In: *Handbook of Theory and Research for the Sociology of Education*, edited by John Richardson, 241–258. New York: Greenwood.

Bourdieu, Pierre, and Loïc Wacquant. 1992. *Invitation to Reflexive Sociology*. Cambridge: Polity.

Bourdieu, Pierre. [1998] 2002. *Masculine Domination*. Stanford: Stanford University Press.

Boyd, John. 1969. "The Algebra of Group Kinship." *Journal of Mathematical Psychology* 6 (1): 139–167.

Brashears, Matthew. 2008. "Gender and Homophily: Differences in Male and Female Association in Blau Space." *Social Science Research* 37 (2): 400–415.

Brashears, Matthew. 2015. "A Longitudinal Analysis of Gendered Association Patterns: Homophily and Social Distance in the General Social Survey." *Journal of Social Structure* 16 (3): 1–26.

Bratman, Michael. 1992. "Shared Cooperative Activity." *The Philosophical Review* 101 (2): 327–341.

Breiger, Ronald. 1974. "The Duality of Persons and Groups." *Social Forces* 53 (2): 181–190.

Breiger, Ronald. 1979. "Toward an Operational Theory of Community Elite Structures." *Quality and Quantity* 13 (1): 21–57.

Breiger, Ronald. 2000. "A Tool Kit for Practice Theory." *Poetics* 27 (2–3): 91–115.

Breiger, Ronald. 2004. "The Analysis of Social Networks." In: *Handbook of Data Analysis*, edited by Melissa Hardy and Alan Bryman, 505–526. London: Sage.

Breiger, Ronald. 2009. "On the Duality of Cases and Variables: Correspondence Analysis (CA) and Qualitative Comparative Analysis (QCA)." In: *The SAGE Handbook of Case-Based Methods*, edited by David Byrne and Charles Ragin, 243–259. London: Sage.

Breiger, Ronald. 2010. "Dualities of Culture and Structure: Seeing Through Cultural Holes." In: *Relationale Soziologie*, edited by Jan Fuhse and Sophie Mützel, 37–47. Wiesbaden: VS-Verlag.

Breiger, Ronald, Robin Wagner-Pacifici, and John Mohr. 2018. "Capturing Distinctions While Mining Text Data: Toward Low-Tech Formalization for Text Analysis." *Poetics* 68: 104–119.

Brint, Steven. 1992. "Hidden Meanings: Cultural Content and Context in Harrison White's Structural Sociology." *Sociological Theory* 10 (2): 194–208.

Brown, Penelope, and Stephen Levinson. [1978] 1987. *Politeness: Some Universals in Language Use*. New York: Cambridge University Press.

Brubaker, Rogers. 2004. *Ethnicity Without Groups*. Cambridge, Massachusetts: Harvard University Press.

Bunge, Mario. 1996. *Finding Philosophy in Social Science*. New Haven: Yale University Press.

Burke, Peter. 1997. "An Identity Model for Network Exchange." *American Sociological Review* 62 (1): 134–150.

Burkitt, Ian. 2016. "Relational Agency: Relational Sociology, Agency and Interaction." *European Journal of Social Theory* 19 (3): 322–339.

Burt, Ronald. 1980. "Models of Network Structure." *Annual Review of Sociology* 6: 79–141.

Burt, Ronald. 1982. *Toward a Structural Theory of Action*. New York: Academic Press.

Burt, Ronald. 1984. "Network Items and the General Social Survey." *Social Networks* 6 (4): 293–339.

Burt, Ronald. 1992. *Structural Holes: The Social Structure of Competition*. Cambridge, Massachusetts: Harvard University Press.

Burt, Ronald. 1998. "The Gender of Social Capital." *Rationality and Society* 10 (1): 5–46.

Burt, Ronald. 2004. "Structural Holes and Good Ideas." *American Journal of Sociology* 110 (2): 349–399.

Butler, Judith. [1990] 2007. *Gender Trouble*. New York: Routledge.

Butts, Carter. 2008. "A Relational Event Framework for Social Action." *Sociological Methodology* 38 (1): 155–200.

Callon, Michel, Jean-Pierre Courtial, William Turner, and Serge Bauin. 1983. "From Translations to Problematic Networks: An Introduction to Co-Word Analysis." *Social Science Information* 22 (2): 191–235.

Cancian, Francesca. 1987. *Love in America*. New York: Cambridge University Press.

Carley, Kathleen. 1986. "An Approach for Relating Social Structure to Cognitive Structure." *Journal of Mathematical Sociology* 12 (2): 137–189.

Carley, Kathleen. 1991. "A Theory of Group Stability." *American Sociological Review* 56 (3): 331–354.

Carley, Kathleen. 1994. "Extracting Culture Through Textual Analysis." *Poetics* 22 (4): 291–312.

Cartwright, Dorwin, and Frank Harary. 1956. "Structural Balance: A Generalization of Heider's Theory." *Psychological Review* 63 (5): 277–292.

Cassirer, Ernst. [1910] 1923. *Function and Substance*. Chicago: Open Court.

Castells, Manuel. [1997] 2010. *The Power of Identity*. Chichester: Wiley-Blackwell.

Castells, Manuel. 2010. *The Rise of the Network Society*. 2nd edition. Chichester: Wiley-Blackwell.

Chafetz, Janet Saltzman. 1990. *Gender Equity: An Integrated Theory of Stability and Change*. Newbury Park: Sage.

Chase, Ivan. 1980. "Social Process and Hierarchy Formation in Small Groups: A Comparative Perspective." *American Sociological Review* 45 (6): 905–924.

Christakis, Nicholas, and James Fowler. 2009. *Connected*. New York: Little, Brown and Company.

Claessens, Dieter. 1963. "Rolle und Verantwortung." *Soziale Welt* 14 (1): 1–13.

Coleman, James. 1988. "Social Capital in the Creation of Human Capital." *American Journal of Sociology* 94 (Supplement): S95–S120.

Coleman, James. 1990a. *Foundations of Social Theory*. Cambridge, Massachusetts: Belknap.

Coleman, James. 1990b. "Rational Action, Social Networks, and the Emergence of Norms." In: *Structures of Power and Constraint*, edited by Craig Calhoun, Marshall Meyer, and Richard Scott, 91–113. Cambridge: Cambridge University Press.

Collins, Randall. 1971. "A Conflict Theory of Sexual Stratification." *Social Problems* 19 (1): 3–21.

Collins, Randall. [1975] 2009. *Conflict Sociology*. Boulder: Paradigm.

Collins, Randall. 1991. "Historical Change and the Ritual Production of Gender." In: *Macro-Micro Linkages in Sociology*, edited by Joan Huber, 109–120. Newbury Park: Sage.

Collins, Randall. 1998. *The Sociology of Philosophies: A Global Theory of Intellectual Change*. Cambridge, Massachusetts: Belknap.

Collins, Randall. 2004. *Interaction Ritual Chains*. Princeton: Princeton University Press.

Collins, Randall, Janet Saltzman Chafetz, Rae Lesser Blumberg, Scott Coltrane, and Jonathan Turner. 1993. "Toward an Integrated Theory of Gender Stratification." *Sociological Perspectives* 36 (3): 185–216.

Connell, R.W. 1987. *Gender and Power*. Palo Alto: Stanford University Press.

Connell, R.W. [1995] 2005. *Masculinities*. Berkeley: University of California Press.

Cook, Karen, and Richard Emerson. 1978. "Power, Equity and Commitment in Exchange Networks." *American Sociological Review* 43 (5): 721–739.

Cooley, Charles Horton. [1902] 1964. *Human Nature and the Social Order*. New York: Schocken.

Cooley, Charles Horton. [1909] 1963. *Social Organization: A Study of the Larger Mind*. New York: Schocken.

Coser, Lewis. 1956. *The Functions of Social Conflict*. London: Routledge.

Cowan, Sarah, and Delia Baldassari. 2018. "'It Could Turn Ugly': Selective Disclosure of Attitudes in Political Discussion Networks." *Social Networks* 52: 1–17.

Crane, Diane. 1972. *Invisible Colleges: Diffusion of Knowledge in Scientific Communities*. Chicago: University of Chicago Press.

Crossley, Nick. 2010. "The Social World of the Network: Combining Qualitative and Quantitative Elements in Social Network Analysis." *Sociologica* 2010 (1): doi: 10.2383/32049.

Crossley, Nick. 2011. *Towards Relational Sociology*. Abingdon: Routledge.

Crossley, Nick. 2015a. *Networks of Sound, Style and Subversion: The Punk and Post-Punk Worlds of Manchester, London, Liverpool and Sheffield, 1975–80.* Manchester: Manchester University Press.

Crossley, Nick. 2015b. "Relational Sociology and Culture: A Preliminary Framework." *International Review of Sociology* 25 (1): 65–85.

Cruickshank, Justin. 2010. "Knowing Social Reality: A Critique of Bhaskar and Archer's Attempt to Derive a Social Ontology from Lay Knowledge." *Philosophy of the Social Sciences* 40 (4): 579–602.

Currarini, Sergio, Matthew Jackson, and Paolo Pin. 2010. "Identifying the Roles of Race-Based Choice and Chance in High School Friendship Network Formation." *Proceedings of the National Academy of Sciences* 107 (11): 4857–4861.

Dahrendorf, Ralf. 1968. *Essays in the Theory of Society.* Stanford: Stanford University Press.

de Beauvoir, Simone. [1949] 1976a. *Le deuxième sexe I.* Paris: Gallimard.

de Beauvoir, Simone. [1949] 1976b. *Le deuxième sexe II.* Paris: Gallimard.

Decker, Scott, and David Curry. 2002. "Gangs, Gang Homicides, and Gang Loyalty: Organized Crimes or Disorganized Criminals." *Journal of Criminal Justice* 30 (4): 343–352.

de Nooy, Wouter. 2003. "Fields and Networks: Correspondence Analysis and Social Network Analysis in the Framework of Field Theory." *Poetics* 31 (5–6): 305–327.

de Nooy, Wouter. 2009. "Formalizing Symbolic Interactionism." *Methodological Innovations Online* 4 (1): 39–52.

de Nooy, Wouter. 2011. "Networks of Action and Events Over Time. A Multilevel Discrete-Time Event History Model for Longitudinal Network Data." *Social Networks* 33 (1): 31–40.

de Nooy, Wouter. 2015. "Structure from Interaction Events." *Big Data & Society* 2 (2): doi: 10.1177/2053951715603732.

de Nooy, Wouter, and Jan Kleinnijenhuis. 2013. "Polarization in the Media During an Election Campaign: A Dynamic Network Model Predicting Support and Attack Among Political Actors." *Political Communication* 30 (1): 117–138.

Denzin, Norman. 1970. "Rules of Conduct and the Study of Deviant Behavior: Some Notes on the Social Relationship." In: *Social Relationships*, edited by George McCall, Michal McCall, Norman Denzin, Gerald Suttles, and Suzanne Kurth, 62–94. Chicago: Aldine.

Dépelteau, François. 2008. "Relational Thinking: A Critique of Co-Deterministic Theories of Structure and Agency." *Sociological Theory* 26 (1): 51–73.

Dépelteau, François. 2015. "Relational Sociology, Pragmatism, Transactions and Social Fields." *International Review of Sociology* 25 (1): 45–64.

Dépelteau, François, ed. 2018. *Palgrave Handbook of Relational Sociology.* New York: Palgrave Macmillan.

Dépelteau, François, Jan Fuhse, Daniel Silver, Prabhu Guptara, Emily Erikson, Normand Carpentier, and Norman Gabriel. 2015. "Invitation to an Ongoing Experiment: Discussing What Relational Sociology Is." *THEORY: International Sociological Association: The Newsletter of the Research Committee on Sociological Theory*, Summer 2015, https://www.academia.edu/13964142/Theory_Summer_2015.

Dépelteau, François, and Christopher Powell, eds. 2013. *Applying Relational Sociology.* New York: Palgrave Macmillan.

de Saussure, Ferdinand. [1916] 2013. *Course in General Linguistics.* London: Bloomsbury.

Desmond, Matthew. 2014. "Relational Ethnography." *Theory & Society* 43 (5): 547–579.

Dewey, John, and Arthur Bentley. [1949] 1973. "Knowing and the Known." In: *Useful Procedures of Inquiry*, edited by Rollo Handy and E.C. Harwood, 97–209. Great Barrington: Behavioral Research Council.

Diani, Mario. 2015. *The Cement of Civil Society*. New York: Cambridge University Press.

Diani, Mario, and Doug McAdam, eds. 2003. *Social Movements and Networks*. New York: Oxford University Press.

Diehl, David. 2019. "Language and Interaction: Applying Sociolinguistics to Social Network Analysis." *Quality & Quantity* 53 (2): 757–774.

DiMaggio, Paul. 1979. "Review Essay: On Pierre Bourdieu." *American Journal of Sociology* 84 (6): 1460–1474.

DiMaggio, Paul. 1986. "Structural Analysis of Organizational Fields: A Blockmodel Approach." *Research in Organizational Behavior* 8: 335–370.

DiMaggio, Paul. 1991. "The Micro-Macro Dilemma in Organizational Research: Implications of Role-Systems Theory." In: *Macro-Micro Linkages in Sociology*, edited by Joan Huber, 76–98. Newbury Park: Sage.

DiMaggio, Paul. 1992. "Nadel's Paradox Revisited: Relational and Cultural Aspects of Organizational Culture." In: *Networks and Organizations*, edited by Nitin Nohria and Robert Eccles, 118–142. Boston: Harvard Business School Press.

DiMaggio, Paul. 1997. "Culture and Cognition." *Annual Review of Sociology* 23: 263–287.

DiMaggio, Paul. 2011. "Cultural Networks." In: *The Sage Handbook of Social Network Analysis*, edited by John Scott and Peter Carrington, 286–301. London: Sage.

DiMaggio, Paul, Manish Ng, and David Blei. 2013. "Exploiting Affinities Between Topic Modeling and the Sociological Perspective on Culture: Application to Newspaper Coverage of U.S. Government Arts Funding." *Poetics* 41 (6): 570–606.

DiMaggio, Paul, and Walter Powell. 1983. "The Iron Cage Revisited: Institutional Isomorphism and Collective Rationality in Organizational Fields." *American Sociological Review* 48 (2): 147–160.

Domínguez, Silvia, and Betina Hollstein, eds. 2014. *Mixed Methods Social Network Research*. Cambridge: Cambridge University Press.

Donati, Pierpaolo. 1983. *Introduzione alle sociologia relazionale*. Milano: FrancoAngeli.

Donati, Pierpaolo. 2011. *Relational Sociology: A New Paradigm for the Social Sciences*. New York: Routledge.

Donati, Pierpaolo. 2013. *Sociologia della relazione*. Bologna: il Mulino.

Donati, Pierpaolo. 2015. "Manifesto for a Critical Realist Relational Sociology." *International Review of Sociology* 25 (1): 86–109.

Donati, Pierpaolo, and Margaret Archer. 2015. *The Relational Subject*. Cambridge: Cambridge University Press.

DuBois, Christopher, Carter Butts, Daniel McFarland, and Padhraic Smyth. 2013. "Hierarchical Models for Relational Event Sequences," *Journal of Mathematical Psychology* 57 (6): 297–309.

Duck, Steve. 1990. "Relationships as Unfinished Business: Out of the Frying Pan and into the 1990s." *Journal of Social and Personal Relationships* 7 (1): 5–28.

Eco, Umberto. [1980] 1983. *The Name of the Rose*. New York: Harcourt.

Eisenstadt, Shmuel, and Luis Roniger. 1980. "Patron-Client Relations as a Model of Structuring Social Exchange." *Comparative Studies in Society and History* 22 (1): 42–77.

Elias, Norbert. [1969] 2006. *The Court Society*. Dublin: University College Dublin Press.

Elias, Norbert. [1970] 2012. *What Is Sociology?* Dublin: University College Dublin Press.

Elias, Norbert. 1974. "Towards a Theory of Communities." In: *The Sociology of Community*, edited by Colin Bell and Howard Newby, ix–xli. London: Cass.

Elias, Norbert. 1987. "The Changing Balance of Power Between the Sexes—A Process-Sociological Study: The Example of the Ancient Roman State." *Theory, Culture & Society* 4 (2–3): 287–316.

Elias, Norbert, and John Scotson. [1965] 1994. *The Established and the Outsiders*. London: Sage.

Eliasoph, Nina, and Paul Lichterman. 2003. "Culture in Interaction." *American Journal of Sociology* 108 (4): 735–794.

Ellwardt, Lea, Giuseppe Labianca, and Rafael Wittek. 2012. "Who Are the Objects of Positive and Negative Gossip at Work? A Social Network Perspective on Workplace Gossip." *Social Networks* 34 (2): 193–205.

Emerson, Richard. 1962. "Power-Dependence Relations." *American Sociological Review* 27 (1): 31–41.

Emerson, Richard. 1972. "Exchange Theory: Part II: Exchange Relations and Networks." In: *Sociological Theories in Progress*, Volume 2, edited by Joseph Berger, Morris Zelditch, and Bo Anderson, 58–87. Boston: Houghton Mifflin.

Emerson, Richard. 1976. "Social Exchange Theory." *Annual Review of Sociology* 2: 335–362.

Emirbayer, Mustafa. 1997. "Manifesto for a Relational Sociology." *American Journal of Sociology* 103 (2): 281–317.

Emirbayer, Mustafa, and Matthew Desmond. 2015. *The Racial Order*. Chicago: The University of Chicago Press.

Emirbayer, Mustafa, and Jeff Goodwin. 1994. "Network Analysis, Culture, and the Problem of Agency." *American Journal of Sociology* 99 (6): 1411–1454.

Emirbayer, Mustafa, and Victoria Johnson. 2008. "Bourdieu and Organizational Analysis." *Theory and Society* 37 (1): 1–44.

Emirbayer, Mustafa, and Ann Mische. 1998. "What Is Agency?" *American Journal of Sociology* 103 (4): 962–1023.

Epstein, A.L. [1961] 1969. "The Network and Urban Social Organization." In: *Social Networks in Urban Situations*, edited by J. Clyde Mitchell, 77–116. Manchester: Manchester University Press.

Epstein, Cynthia Fuchs. 1988. *Deceptive Distinctions: Sex, Gender, and the Social Order*. New Haven: Yale University Press.

Erickson, Bonnie. 1988. "The Relational Basis of Attitudes." In: *Social Structures: A Network Approach*, edited by Barry Wellman and Stephen Berkowitz, 99–121. Cambridge: Cambridge University Press.

Erickson, Bonnie. 1996. "Culture, Class, and Connections." *American Journal of Sociology* 102 (1): 217–251.

Erikson, Emily. 2013. "Formalist and Relationalist Theory in Social Network Analysis." *Sociological Theory* 31 (3): 219–242.

Erikson, Emily, and Peter Bearman. 2006. "Malfeasance and the Foundations for Global Trade: The Structure of English Trade in the East Indies, 1601–1833." *American Journal of Sociology* 112 (1): 195–230.

Erikson, Emily, and Nicholas Occhiuto. 2017. "Social Networks and Macrosocial Change." *Annual Review of Sociology* 43: 229–248.

Esser, Hartmut. 2004. "Does the 'New' Immigration Require a 'New' Theory of Intergenerational Integration?" *International Migration Review* 38 (3): 1126–1159.

Evans, James, and Pedro Aceves. 2016. "Machine Translation: Mining Text for Social Theory." *Annual Review of Sociology* 42: 21–50.

Feld, Scott. 1981. "The Focused Organization of Social Ties." *American Journal of Sociology* 86 (5): 1015–1035.

Feyerabend, Paul. 1962. "Explanation, Reduction, and Empiricism." *Minnesota Studies in the Philosophy of Science* 3: 28–97.

Fincham, Frank, and Ross May. 2017. "Infidelity in Romantic Relationships." *Current Opinion in Psychology* 13: 70–74.

Fine, Gary Alan. 1979. "Small Groups and Culture Creation: The Idioculture of Little League Baseball Teams." *American Sociological Review* 44 (5): 733–745.

Fine, Gary Alan. 2012. *Tiny Publics: A Theory of Group Action and Culture.* New York: Russell Sage Foundation.

Fine, Gary Alan. 2021. *The Hinge: Civil Society, Group Cultures, and the Power of Local Commitments.* Chicago: University of Chicago Press.

Fine, Gary Alan, and Sherryl Kleinman. 1979. "Rethinking Subculture: An Interactionist Analysis." *American Journal of Sociology* 85 (1): 1–20.

Fine, Gary Alan, and Sherryl Kleinman. 1983. "Network and Meaning: An Interactionist Approach to Structure." *Symbolic Interaction* 6 (1): 97–110.

Fine, Gary Alan, Jeffrey Stitt, and Michael Finch. 1984. "Couple Tie-Signs and Interpersonal Threat: A Field Experiment." *Social Psychology Quarterly* 47 (3): 282–286.

Fischer, Claude. 1982a. *To Dwell Among Friends: Personal Networks in Town and City.* Chicago: University of Chicago Press.

Fischer, Claude. 1982b. "What Do We Mean By 'Friend'? An Inductive Study." *Social Networks* 3 (4): 287–306.

Fischer, Claude. 2009. "The 2004 GSS Finding of Shrunken Social Networks: An Artifact?" *American Sociological Review* 74 (4): 657–669.

Fischer, Claude, and Stacey Oliker. 1983. "A Research Note on Friendship, Gender, and the Life Cycle." *Social Forces* 62 (1): 124–133.

Fong, Eric, and Wsevolod Isajiw. 2000. "Determinants of Friendship Choices in Multiethnic Society." *Sociological Forum* 15 (2): 249–271.

Fontdevila, Jorge. 2010. "Indexes, Power, and Netdoms: A Multidimensional Model of Language in Social Action." *Poetics* 38 (6): 587–609.

Fontdevila, Jorge, Pilar Opazo, and Harrison White. 2011. "Order at the Edge of Chaos: Meanings from Netdom Switchings Across Functional Systems." *Sociological Theory* 29 (3): 178–198.

Foucault, Michel. 1969. *L'archéologie du savoir.* Paris: Gallimard.

Foucault, Michel. [1969] 1972. *The Archeology of Knowledge.* London: Tavistock.

Foucault, Michel. [1976] 1978. *The History of Sexuality: Volume 1: An Introduction.* New York: Pantheon.

Foucault, Michel. 1998. *Aesthetics, Method, and Epistemology.* New York: The New Press.

Freeman, Linton. 1992. "The Sociological Concept of 'Group': An Empirical Test of Two Models." *American Journal of Sociology* 98 (1): 152–166.

Freeman, Linton. 2004. *The Development of Social Network Analysis.* Vancouver: Empirical Press.

Friedkin, Noah. 1980. "A Test of Structural Features of Granovetter's Strength of Weak Ties Theory." *Social Networks* 2 (4): 411–422.

Friedland, Roger, and Robert Alford. 1991. "Bringing Society Back In: Symbols, Practices, and Institutional Contradictions." In: *The New Institutionalism in Organizational*

Analysis, edited by Paul DiMaggio and Walter Powell, 232–263. Chicago: University of Chicago Press.

Friedland, Roger, John Mohr, Henk Roose, and Paolo Gardinali. 2014. "The Institutional Logics of Love: Measuring Intimate Life." *Theory & Society* 43 (3/4): 333–370.

Fuchs, Stephan. 2001a. *Against Essentialism: A Theory of Culture and Society*. Cambridge, Massachusetts: Harvard University Press.

Fuchs, Stephan. 2001b. "Beyond Agency." *Sociological Theory* 19 (1): 24–40.

Fuhse, Jan. 2006. "Gruppe und Netzwerk—eine begriffsgeschichtliche Rekonstruktion." *Berliner Journal für Soziologie* 16 (2): 245–263.

Fuhse, Jan. 2008. *Ethnizität, Akkulturation und persönliche Netzwerke von italienischen Migranten*. Leverkusen: Barbara Budrich.

Fuhse, Jan. 2012. "Embedding the Stranger: Ethnic Categories and Cultural Differences in Social Networks." *Journal of Intercultural Studies* 33 (2012): 639–655.

Fuhse, Jan. 2013. "Social Relationships Between Communication, Network Structure, and Culture." In: *Applying Relational Sociology*, edited by François Dépelteau and Christopher Powell, 181–206. New York: Palgrave Macmillan.

Fuhse, Jan. 2014. "Kulturelle Differenz und Kommunikation in Netzwerken." In: *Neue Impulse für die soziologische Kommunikationstheorie*, edited by Thomas Malsch and Marco Schmitt, 91–120. Wiesbaden: VS.

Fuhse, Jan. 2015a. "Culture and Social Networks." In: *Emerging Trends in the Social and Behavioral Sciences*, edited by Robert Scott and Stephen Kosslyn. New York: Wiley. doi: 10.1002/9781118900772.etrds0066.

Fuhse, Jan. 2015b. "Theorizing Social Networks: Relational Sociology of and Around Harrison White." *International Review of Sociology* 25 (1): 15–44.

Fuhse, Jan. 2018a. "Deconstructing and Reconstructing Social Networks." In: *The Palgrave Handbook of Relational Sociology*, edited by François Dépelteau, 457–479. New York: Palgrave.

Fuhse, Jan. 2018b. "New Media and Socio-Cultural Formations." *Cybernetics & Human Knowing* 25 (4): 73–96.

Fuhse, Jan. 2020a. "Relational Sociology of the Scientific Field: Communication, Identities, and Field Relations." *Digithum: A Relational Perspective on Culture and Society* 26, 1–14. doi: 10.7238/d.v0i26.374144.

Fuhse, Jan. 2020b. "The Field of Relational Sociology." *Digithum: A Relational Perspective on Culture and Society* 26, 1–10. doi: 10.7238/d.v0i26.374145.

Fuhse, Jan. 2020c. "Theories of Social Networks." In: *Handbook of Social Networks*, edited by Ryan Light and James Moody, 34–49. New York: Oxford University Press.

Fuhse, Jan, and Sophie Mützel. 2011. "Tackling Connections, Structure, and Meaning in Networks: Quantitative and Qualitative Methods in Sociological Network Research." *Quality & Quantity* 45 (5): 1067–1089.

Fuhse, Jan, Oscar Stuhler, Jan Riebling, and John Levi Martin. 2020. "Relating Social and Symbolic Relations in Quantitative Text Analysis. A Study of Parliamentary Discourse in the Weimar Republic." *Poetics* 78: Article 101363.

Galtung, Johan. 1959. "Expectations and Interaction Processes." *Inquiry* 2 (1–4): 213–234.

Gans, Herbert. 1979. "Symbolic Ethnicity: The Future of Ethnic Groups and Cultures in America." In: *On the Making of Americans*, edited by Herbert Gans, Nathan Glazer, Joseph Gusfield, and Christopher Jencks, 193–220. Philadelphia: University of Pennsylvania Press.

Garfinkel, Harold. [1967] 1984. *Studies in Ethnomethodology*. Cambridge: Polity.

Geertz, Clifford. 1973. *The Interpretation of Cultures*. London: Fontana.

Gehlen, Arnold. [1940] 1988. *Man: His Nature and Place in the World*. New York: Columbia University Press.

Gehlen, Arnold. 1956. *Urmensch und Spätkultur*. Bonn: Athenäum.

Gibson, David. 2000. "Seizing the Moment: The Problem of Conversational Agency." *Sociological Theory* 18 (3): 368–382.

Gibson, David. 2003. "Participation Shifts: Order and Differentiation in Group Conversation." *Social Forces* 81 (4): 1335–1381.

Gibson, David. 2005. "Taking Turns and Talking Ties: Networks and Conversational Interaction." *American Journal of Sociology* 110 (6): 1561–1597.

Giddens, Anthony. 1992. *The Transformation of Intimacy*. Cambridge: Polity.

Giere, Ronald. 2006a. *Scientific Perspectivism*. Chicago: University of Chicago Press.

Giere, Ronald. 2006b. "Perspectival Pluralism." In: *Scientific Pluralism*, edited by Stephen Kellert, Helen Longino, and Kenneth Waters, 26–41. Minneapolis: University of Minnesota Press.

Gilbert, Margaret. 1990. "Walking Together: A Paradigmatic Social Phenomena." *Midwest Studies in Philosophy* 15 (1): 1–14.

Gilligan, Carol. [1982] 2016. *In a Different Voice*. Cambridge, Massachusetts: Harvard University Press.

Godart, Frédéric, and Harrison White. 2010. "Switchings Under Uncertainty. The Coming and Becoming of Meanings." *Poetics* 38 (6): 567–586.

Goffman, Erving. [1955] 1967. "On Face-Work." In: Erving Goffman: *Interaction Ritual*, 5–45. Harmondsworth: Penguin.

Goffman, Erving. [1959] 1990. *The Presentation of Self in Everyday Life*. London: Penguin.

Goffman, Erving. [1974] 1986. *Frame Analysis: An Essay on the Organization of Experience*. Boston: Northeastern University Press.

Goffman, Erving. 1977. "The Arrangement Between the Sexes." *Theory and Society* 4 (3): 301–331.

Goffman, Erving. 1981. *Forms of Talk*. Philadelphia: University of Pennsylvania Press.

Goffman, Erving. 1983. "The Interaction Order." *American Sociological Review* 48 (1): 1–17.

Gondal, Neha, and Paul McLean. 2013. "What Makes a Network Go Round: Exploring the Structure of a Strong Component with Exponential Random Graph Models." *Social Networks* 35 (4): 499–513.

Goodwin, Jeff. 2006. "A Theory of Categorical Terrorism." *Social Forces* 84 (4): 2027–2046.

Goodwin, Marjorie Harness. 1990. "Tactical Uses of Stories: Participation Frameworks Within Girls' and Boys' Disputes." *Discourse Processes* 13 (1): 33–71.

Gordon, Milton. 1964. *Assimilation in American Life: The Role of Race, Religion, and National Origins*. New York: Oxford University Press.

Gould, Roger. 1995. *Insurgent Identities: Class, Community, and Protest in Paris from 1848 to the Commune*. Chicago: University of Chicago Press.

Granovetter, Mark. 1973. "The Strength of Weak Ties." *American Journal of Sociology* 78 (6): 1360–1380.

Granovetter, Mark. 2007. "Introduction for the French Reader." *Sociologica* 2/2007. doi: 10.2383/24767.

Green, Adam Isaiah. 2011. "Playing the (Sexual) Field: The Interactional Basis of Systems of Sexual Stratification." *Social Psychology Quarterly* 74 (3): 244–266.

Green, Adam Isaiah, ed. 2014. *Sexual Fields*. Chicago: University of Chicago Press.

Griffith, Belver, and Nicholas Mullins. 1972. "Coherent Social Groups in Scientific Change: 'Invisible Colleges' May Be Consistent Throughout Science." *Science* 177 (4053): 959–964.

Gumperz, John. 1982. *Discourse Strategies*. Cambridge: Cambridge University Press.

Gumplowicz, Ludwig. [1885] 1999. *The Outlines of Sociology*. Kitchener: Batoche.

Gusfield, Joseph. [1963] 1986. *Symbolic Crusade: Status Politics and the American Temperance Movement*. Urbana: University of Illinois Press.

Guy, Jean-Sébastien. 2018. "Is Niklas Luhmann a Relational Sociologist?" In: *Palgrave Handbook of Relational Sociology*, edited by François Dépelteau, 289–304. New York: Palgrave Macmillan.

Habermas, Jürgen. [1992] 1996. *Between Facts and Norms*. Cambridge: Polity Press.

Hallinan, Maureen, and Richard Williams. 1989. "Interracial Friendship Choices in Secondary Schools." *American Sociological Review* 54 (1): 67–78.

Hannerz, Ulf. 1992. *Cultural Complexity: Studies in the Social Organization of Meaning*. New York: Columbia University Press.

Hannerz, Ulf. 1996. *Transnational Connections: Culture, People, Places*. London: Routledge.

Hanson, Norwood. 1958. *Patterns of Discovery*. Cambridge: Cambridge University Press.

Hardwick, Susan. 2003. "Migration, Embedded Networks and Social Capital." *International Journal of Population Geography* 9 (2): 163–179.

Hebdige, Dick. 1979. *Subculture: The Meaning of Style*. London: Routledge.

Hedström, Peter. 2005. *Dissecting the Social: On the Principles of Analytical Sociology*. Cambridge: Cambridge University Press.

Heritage, John, and Geoffrey Raymond. 2005. "The Terms of Agreement: Indexing Epistemic Authority and Subordination in Talk-in-Interaction." *Social Psychology Quarterly* 68 (1): 15–38.

Hesse, Mary. 1962. *Forces and Fields*. Mineola, New York: Dover.

Hesse, Mary. 1974. *The Structure of Scientific Inference*. London: Macmillan.

Hinde, Robert. 1976. "Interactions, Relationships and Social Structure." *Man* 11 (1): 1–17.

Hoffmann, Bruce. 2006. *Inside Terrorism*. New York: Columbia University Press.

Hogg, Michael, and Dominic Abrams. 1988. *Social Identifications: A Social Psychology of Intergroup Relations and Group Processes*. London: Routledge.

Holland, Paul, and Samuel Leinhardt. 1977. "Social Structure as a Network Process." *Zeitschrift für Soziologie* 6 (4): 386–402.

Holzer, Boris. 2006. *Netzwerke*. Bielefeld: Transcript Verlag.

Holzer, Boris. 2008. "Netzwerke und Systeme: Zum Verhältnis von Vernetzung und Differenzierung." In: *Netzwerkanalyse und Netzwerktheorie*, edited by Christian Stegbauer, 155–164. Wiesbaden: VS.

Homans, George Caspar. 1950. *The Human Group*. New York: Harcourt, Brace and Company.

Homans, George Caspar. 1961. *Social Behavior: Its Elementary Forms*. New York: Harcourt, Brace & World.

Homans, George Caspar. 1964. "Bringing Men Back In." *American Sociological Review* 29 (5): 809–818.

Hopper, Robert, Mark Knapp, and Lorel Scott. 1981. "Couples' Personal Idioms: Exploring Intimate Talk." *Journal of Communication* 31 (1): 23–33.

Horowitz, Donald. 1985. *Ethnic Groups in Conflict*. Berkeley: University of California Press.

Ibarra, Herminia. 1992. "Homophily and Differential Returns: Sex Differences in Network Structure and Access in an Advertising Firm." *Administrative Science Quarterly* 37 (3): 422–447.

Ikegami, Eiko. 2005. *Bonds of Civility: Aesthetic Networks and the Political Origins of Japanese Culture*. New York: Cambridge University Press.

Illouz, Eva. 1998. "The Lost Innocence of Love: Romance as a Postmodern Condition." *Theory, Culture & Society* 15 (3–4): 161–186.

Illouz, Eva. 2012. *Why Love Hurts*. Cambridge: Polity.

Irigaray, Luce. [1977] 1985. *The Sex Which Is Not One*. Ithaca: Cornell University Press.

Jackson, Patrick Thaddeus, and Daniel Nexon. 1999. "Relations Before States: Substance, Process and the Study of World Politics." *European Journal of International Relations* 5 (3): 291–332.

Jamieson, Lynn. 1998. *Intimacy: Personal Relationships in Modern Societies*. Cambridge: Polity.

Kao, Grace, and Kara Joyner. 2006. "Do Hispanic and Asian Adolescents Practice Panethnicity in Friendship Choices?" *Social Science Quarterly* 87 (5): 972–992.

Kaufmann, Jean-Claude. 1992. *La trame conjugale: Analyse du couple par son linge*. Paris: Nathan.

Kaufmann, Jean-Claude. 1993. *Sociologie du couple*. Paris: Presses Universitaires de France.

Kaufmann, Jean-Claude. [1999] 2008. *The Single Woman and the Fairytale Prince*. Cambridge: Polity.

Kessler, Suzanne, and Wendy McKenna. 1978. *Gender: An Ethnomethodological Approach*. Chicago: University of Chicago Press.

Ketokivi, Kaisa. 2012. "The Intimate Couple, Family and the Relational Organization of Close Relationships." *Sociology* 46 (3): 473–489.

Kitts, James. 2014. "Beyond Networks in Structural Theories of Exchange: Promises from Computational Social Science." *Advances in Group Processes* 31: 263–298.

Kitts, James, Alessandro Lomi, Daniele Mascia, Francesca Pallotti, and Eric Quintane. 2017. "Investigating the Temporal Dynamics of Interorganizational Exchange: Patient Transfers Among Italian Hospitals." *American Journal of Sociology* 123 (3): 850–910.

Kitts, James, and Eric Quintane. 2020. "Rethinking Social Networks in the Era of Computational Social Science." In: *The Oxford Handbook of Social Networks*, edited by Ryan Light and James Moody, 71–97. New York, Oxford University Press.

Klapp, Orrin. 1969. *Collective Search for Identity*. New York: Holt, Rinehart and Winston.

Klein, Malcolm. 1995. *The American Street Gang*. New York: Oxford University Press.

Kossinets, Gueorgi, and Duncan Watts. 2009. "Origins of Homophily in an Evolving Social Network." *American Journal of Sociology* 115 (2): 405–450.

Krackhardt, David. 1999. "The Ties That Torture: Simmelian Tie Analysis in Organizations." *Research in the Sociology of Organizations* 16: 183–210.

Kühl, Stefan. 2020. "Groups, Organizations, Families and Movements: The Sociology of Social Systems Between Interaction and Society." *Systems Research & Behavioral Science* 37 (3): 496–515.

Lamont, Michèle. 1992. *Money, Morals, & Manners: The Culture of the French and the American Upper-Middle Class*. Chicago: University of Chicago Press.

Lamont, Michèle. 2000. *The Dignity of Working Men: Morality and the Boundaries of Race, Class, and Immigration*. Cambridge, Massachusetts: Harvard University Press.

Lamont, Michèle, and Marcel Fournier, eds. 1992. *Cultivating Differences: Symbolic Boundaries and the Making of Inequality*. Chicago: University of Chicago Press.

Lamont, Michèle, and Virág Molnár. 2002. "The Study of Boundaries Across the Social Sciences." *Annual Review of Sociology* 28: 167–195.

Lash, Scott. 1993. "Pierre Bourdieu: Cultural Economy and Social Change." In: *Bourdieu: Critical Perspectives*, edited by Craig Calhoun, Edward LiPuma, and Moishe Postone, 193–211. Cambridge: Polity.

Latour, Bruno. 1993. *We Have Never Been Modern*. Cambridge, Massachusetts: Harvard University Press.

Latour, Bruno. 2005. *Reassembling the Social*. Oxford: Oxford University Press.

Latour, Bruno. 2012. *Enquête sur les modes d'existence*. Paris: La Découverte.

Latour, Bruno, Pablo Jensen, Tommaso Venturini, Sébastian Grauwin, and Dominique Boullier. 2012. "'The Whole Is Always Smaller than Its Parts'—a Digital Test of Gabriel Tardes' Monads." *British Journal of Sociology* 63 (4): 590–615.

Laumann, Edward. 1979. "Network Analysis in Large Social Systems." In: *Perspectives on Social Network Research*, edited by Paul Holland and Samuel Leinhardt, 379–402. New York: Academic Press.

Lawler, Edward, Shane Thye, and Jeongkoo Yoon. 2008. "Social Exchange and Micro Social Order." *American Sociological Review* 73 (4): 519–542.

Lazarsfeld, Paul, and Robert Merton. 1954. "Friendship as Social Process: A Substantial and Methodological Analysis." In: *Freedom and Control in Modern Society*, edited by Morroe Berger, Theodore Abel, and Charles Page, 18–66. Toronto: Van Nostrand.

Lazega, Emmanuel, and Philippa Pattison. 1999. "Multiplexity, Generalized Exchange and Cooperation in Organizations: A Case Study." *Social Networks* 21 (1): 67–90.

Lazer, David, Alex Pentland, Lada Adamic, Sinan Aral, Albert-László Barabási, Devon Brewer, Nicholas Christakis, Noshir Contractor, James Fowler, Myron Gutmann, Tony Jebara, Gary King, Michael Macy, Deb Roy, and Marshall Van Alstyne. 2009. "Computational Social Science." *Science* 323 (5915): 721–723.

Lee, Monica, and John Levi Martin. 2018. "Doorway to the Dharma of Duality." *Poetics* 68: 18–30.

Leifer, Eric. 1988. "Interaction Preludes to Role Setting: Exploratory Local Action." *American Sociological Review* 53 (6): 865–878.

Lerner, Gene. 1996. "Finding 'Face' in the Preference Structures of Talk-in-Interaction." *Social Psychology Quarterly* 59 (4): 303–321.

Leszczensky, Lars, and Sebastian Pink. 2019. "What Drives Ethnic Homophily? A Relational Approach on How Ethnic Identification Moderates Preferences for Same-Ethnic Friends." *American Sociological Review* 84 (3): 394–419.

Lévi-Strauss, Claude. [1949] 1969. *The Elementary Structures of Kinship*. Boston: Beacon Press.

Lewis, Kevin. 2013. "The Limits of Racial Prejudice." *Proceedings of the National Academy of Sciences* 110 (47): 18814–18819.

Leydesdorff, Loet. 2000. *A Sociological Theory of Communication: The Self-Organization of the Knowledge-Based Society*. Boca Raton: Universal Publishers.

Leydesdorff, Loet. 2007. "Scientific Communication and Cognitive Codification: Social Systems Theory and the Sociology of Scientific Knowledge." *European Journal of Social Theory* 10 (3): 375–388.

Leydesdorff, Loet. 2010. "The Communication of Meaning and the Structuration of Expectations: Giddens' 'Structuration Theory' and Luhmann's 'Self-Organization.'"

Journal of the American Society for Information Science and Technology 61 (10): 2138–2150.

Lieberson, Stanley. 1981. *Language Diversity and Language Contact*. Stanford: Stanford University Press.

Light, Ryan. 2014. "From Words to Networks and Back: Digital Text, Computational Social Science, and the Case of Presidential Inaugural Addresses." *Social Currents* 1 (2): 111–129.

Lin, Nan. 2001. *Social Capital*. Cambridge: Cambridge University Press.

Lin, Nan, Walter Ensel, and John Vaughn. 1981. "Social Resources and Strength of Ties: Structural Factors in Occupational Status Attainment." *American Sociological Review* 46 (4): 393–405.

Lindemann, Gesa. 2009. *Das Soziale von seinen Grenzen her denken*. Weilerswist: Velbrück.

Lindenberg, Siegwart, and Bruno Frey. 1993. "Alternatives, Frames, and Relative Prices: A Broader View of Rational Choice Theory." *Acta Sociologica* 36 (3): 191–205.

Linton, Ralph. 1936. *The Study of Man*. Oxford: Appleton-Century.

Lizardo, Omar. 2006. "How Cultural Tastes Shape Personal Networks." *American Sociological Review* 71 (5): 778–807.

Locher, Miriam, and Richard Watts. 2005. "Politeness Theory and Relational Work." *Journal of Politeness Research* 1 (1): 9–33.

Lorrain, François. 1975. *Réseaux sociaux et classifications sociales*. Paris: Hermann.

Lorrain, François, and Harrison White. 1971. "Structural Equivalence of Individuals in Social Networks." *Journal of Mathematical Sociology* 1 (1): 49–80.

Louch, Hugh. 2000. "Personal Network Integration: Transitivity and Homophily in Strong-Tie Relations." *Social Networks* 22 (1): 45–64.

Loveman, Mara. 1999. "Is 'Race' Essential?" *American Sociological Review* 64 (6): 891–898.

Lubbers, Miranda. 2003. "Group Composition and Network Structure in School Classes: A Multilevel Application of the *p*∗ Model." *Social Networks* 25 (4): 309–332.

Lubbers, Miranda, José Luis Molina, and Christopher McCarty. 2007. "Personal Networks and Ethnic Identifications." *International Sociology* 22 (6): 721–741.

Luhmann, Niklas. 1970. "Institutionalisierung—Funktion und Mechanismus im sozialen System der Gesellschaft." In: *Zur Theorie der Institution*, edited by Helmut Schelsky, 28–41. Düsseldorf: Bertelsmann.

Luhmann, Niklas. [1982] 1986. *Love as Passion*. Cambridge, Massachusetts: Harvard University Press.

Luhmann, Niklas. [1984] 1995. *Social Systems*. Stanford: Stanford University Press.

Luhmann, Niklas. 1990. *Essays on Self-Reference*. New York: Oxford University Press.

Luhmann, Niklas. 1995a. *Gesellschaftsstruktur und Semantik 4*. Frankfurt am Main: Suhrkamp.

Luhmann, Niklas. 1995b. *Soziologische Aufklärung 6*. Opladen: Westdeutscher Verlag.

Luhmann, Niklas. [1997] 2012. *Theory of Society*, Volume 1. Stanford: Stanford University Press.

Luhmann, Niklas. [1997] 2013. *Theory of Society*, Volume 2. Stanford: Stanford University Press.

Luhmann, Niklas. 2002. *Theories of Distinction*, edited by William Rasch. Stanford: Stanford University Press.

Lüsebrink, Hans-Jürgen. 2008. *Interkulturelle Kommunikation: Interaktion, Fremdwahrnehmung, Kulturtransfer*. Stuttgart: Metzler.

Lusher, Dean, Johan Koskinen, and Garry Robins. 2013. *Exponential Graph Models for Social Networks*. Cambridge: Cambridge University Press.

Maffesoli, Michel. [1988] 1996. *The Time of the Tribes*, London: Sage.

Mark, Noah. 1998. "Beyond Individual Differences: Social Differentiation from First Principles." *American Sociological Review* 63 (3): 309–330.

Mark, Noah. 2003. "Culture and Competition: Homophily and Distancing Explanations for Cultural Niches." *American Sociological Review* 68 (3): 319–345.

Marsden, Peter. 1983. "Restricted Access in Networks and Models of Power." *American Journal of Sociology* 88 (4): 686–717.

Marsden, Peter. 1987. "Core Discussion Networks of Americans." *American Sociological Review* 52 (1): 122–131.

Martin, John Levi. 2000. "What Do Animals Do All Day? The Division of Labor, Class Bodies, and Totemic Thinking in the Popular Imagination." *Poetics* 27 (2–3): 195–231.

Martin, John Levi. 2002. "Power, Authority, and the Constraint of Belief Systems." *American Journal of Sociology* 107 (4): 861–904.

Martin, John Levi. 2003. "What Is Field Theory?" *American Journal of Sociology* 109 (1): 1–49.

Martin, John Levi. 2009. *Social Structures*. Princeton: Princeton University Press.

Martin, John Levi. 2011. *The Explanation of Social Action*. New York: Oxford University Press.

Martin, John Levi, and Matt George. 2006. "Theories of Sexual Stratification: Toward an Analytics of the Sexual Field and a Theory of Sexual Capital." *Sociological Theory* 24 (2): 107–132.

Martin, John Levi, and King-To Yeung. 2003. "The Use of the Conceptual Category of Race in American Sociology." *Sociological Forum* 18 (4): 521–543.

Martin, Patricia Yancey. 2004. "Gender as Social Institution." *Social Forces* 82 (4): 1249–1273.

Marx, Karl. [1852] 2005. *The Eighteenth Brumaire of Louis Bonaparte*. New York: Mondial 2005.

Massey, Douglas. 2007. *Categorically Unequal: The American Stratification System*. New York: Russell Sage Foundation.

Mayer, Adrian. 1966. "The Significance of Quasi-Groups in the Study of Complex Societies." In: *The Social Anthropology of Complex Societies*, edited by Michael Banton, 97–122. London: Tavistock.

Mayer, Philip, and Iona Mayer. 1961. *Townsmen or Tribesmen: Conservatism and the Process of Urbanization in a South African City*. Cape Town: Oxford University Press.

Maynard, Douglas, and Don Zimmerman. 1984. "Topical Talk, Ritual and the Social Organization of Relationships." *Social Psychology Quarterly* 47 (4): 301–316.

McAdam, Doug. 1986. "Recruitment to High-Risk Activism: The Case of Freedom Summer." *American Journal of Sociology* 92 (1): 64–90.

McCall, George. 1970. "The Social Organization of Relationships." In: *Social Relationships*, edited by George McCall, Michal McCall, Norman Denzin, Gerald Suttles, and Suzanne Kurth, 3–34. Chicago: Aldine.

McCall, George, and J.L. Simmons. [1966] 1978. *Identities and Interactions*. New York: Free Press.

McFarland, Daniel. 2001. "Student Resistance: How the Formal and Informal Organization of Classrooms Facilitate Everyday Forms of Student Defiance." *American Journal of Sociology* 107 (3): 612–678.

McFarland, Daniel. 2004. "Resistance as a Social Drama—a Study of Change-Oriented Encounters." *American Journal of Sociology* 109 (6): 1249–1318.

McFarland, Daniel, Dan Jurafsky, and Craig Rawlings. 2013. "Making the Connection: Social Bonding in Courtship Situations." *American Journal of Sociology* 118 (6): 1596–1649.

McFarland, Daniel, Kevin Lewis, and Amir Goldberg. 2015. "Sociology in the Era of Big Data: The Ascent of Forensic Social Science." *The American Sociologist* 47 (1): 12–35.

McFarland, Daniel, James Moody, David Diehl, Jeffrey Smith, and Reuben Thomas. 2014. "Network Ecology and Adolescent Social Structure." *American Sociological Review* 79 (6): 1088–1121.

McKinney, John. 1969. "Typification, Typologies, and Sociological Theory." *Social Forces* 48 (1): 1–12.

McLean, Paul. 1998. "A Frame Analysis of Favor Seeking in the Renaissance: Agency, Networks, and Political Culture." *American Journal of Sociology* 104 (1): 51–91.

McLean, Paul. 2007. *The Art of the Network: Strategic Interaction and Patronage in Renaissance Florence*. Durham: Duke University Press.

McLean, Paul. 2017. *Culture in Networks*. New York: Wiley.

McPherson, Miller, Lynn Smith-Lovin, and Matthew Brashears. 2006. "Social Isolation in America: Changes in Core Discussion Networks over Two Decades." *American Sociological Review* 71 (3): 353–375.

McPherson, Miller, Lynn Smith-Lovin, and James Cook. 2001. "Birds of a Feather: Homophily in Social Networks." *Annual Review of Sociology* 27: 415–444.

Mead, George Herbert. [1934] 1967. *Mind, Self & Society*. Chicago: Chicago University Press.

Mears, Ashley. 2015. "Working for Free in the VIP: Relational Work and the Production of Consent." *American Sociological Review* 80 (6): 1099–1122.

Merton, Robert. 1957. "The Role-Set: Problems in Sociological Theory." *British Journal of Sociology* 8 (2): 106–120.

Merton, Robert. 1958. *Social Theory and Social Structure*. New York: The Free Press.

Merton, Robert. 1968. "The Matthew Effect in Science." *Science* 159 (3810): 56–63.

Meyer, John, John Boli, George Thomas, and Francisco Ramirez. 1997. "World Society and the Nation-State." *American Journal of Sociology* 103 (1): 144–181.

Meyer, John, and Ronald Jepperson. 2000. "The 'Actors' of Modern Society: The Cultural Construction of Social Agency." *Sociological Theory* 18 (1): 100–120.

Meyer, John, and Brian Rowan. 1977. "Institutionalized Organizations: Formal Structure as Myth and Ceremony." *American Journal of Sociology* 83 (2): 340–363.

Michels, Robert. [1911] 2001. *Political Parties*. Kitchener: Batoche.

Mills, C. Wright. 1940. "Situated Actions and Vocabularies of Motive." *American Sociological Review* 5 (6): 904–913.

Mills, C. Wright. 1956. *The Power Elite*. New York: Oxford University Press.

Mills, Theodore. [1967] 1984. *The Sociology of Small Groups*. Englewood Cliffs: Prentice Hall.

Mische, Ann. 2003. "Cross-Talk in Movements: Reconceiving the Culture-Network Link." In: *Social Movements and Networks*, edited by Mario Diani and Doug McAdam, 258–280. New York: Oxford University Press.

Mische, Ann. 2008. *Partisan Publics: Communication and Contention Across Brazilian Youth Activist Networks*. Princeton: Princeton University Press.

Mische, Ann. 2011. "Relational Sociology, Culture, and Agency." In: *The Sage Handbook of Social Network Analysis*, edited by John Scott and Peter Carrington, 80–97. London: Sage.

Mische, Ann, and Harrison White. 1998. "Between Conversation and Situation: Public Switching Dynamics Across Network Domains." *Social Research* 65 (3): 695–724.

Mitchell, J. Clyde. 1969. "The Concept and Use of Social Networks." In: *Social Networks in Urban Situations*, edited by J. Clyde Mitchell, 1–50. Manchester: Manchester University Press.

Mitchell, J. Clyde. 1973. "Networks, Norms and Institutions." In: *Network Analysis*, edited by Jeremy Boissevain and J. Clyde Mitchell, 2–35. Den Haag: Mouton.

Mitchell, J. Clyde. 1974. "Perceptions of Ethnicity and Ethnic Behaviour." In: *Urban Ethnicity*, edited by Abner Cohen, 1–35. London: Tavistock.

Mohr, John. 1994. "Soldiers, Mothers, Tramps, and Others: Discourse Roles in the 1907 New York Charity Discourse." *Poetics* 22 (4): 327–357.

Mohr, John. 1998. "Measuring Meaning Structures." *Annual Review of Sociology* 24: 345–370.

Mohr, John. 2000. "Introduction: Structures, Institutions, and Cultural Analysis." *Poetics* 27 (2–3): 57–68.

Mohr, John. 2011. "Ernst Cassirer: Science, Symbols, and Logics." In: *Sociological Insights of Great Thinkers*, edited by Christofer Edling and Jens Rydgren, 113–122. Santa Barbara: Praeger.

Mohr, John. 2013. "Bourdieu's Relational Method in Theory and in Practice." In: *Conceptualizing Relational Sociology*, edited by Christopher Powell and François Dépelteau, 101–135. New York: Palgrave Macmillan.

Mohr, John, and Vincent Duquenne. 1997. "The Duality of Culture and Practices: Poverty Relief in New York City, 1888–1917." *Theory and Society* 26 (2/3): 305–356.

Mohr John, Robin Wagner-Pacifici, and Ronald Breiger. 2013. "Graphing the Grammar of Motives in U.S. National Security Strategies: Cultural Interpretation, Automated Text Analysis and the Drama of Global politics." *Poetics* 41 (6): 670–700.

Mohr, John, Robin Wagner-Pacifici, and Ronald Breiger. 2015. "Toward a Computational Hermeneutics." *Big Data & Society* 2 (2): 1–15.

Mohr, John, and Harrison White. 2008. "How to Model an Institution." *Theory and Society* 37 (5): 485–512.

Mollenhorst, Gerald, Beate Volker, and Henk Flap. 2014. "Changes in Personal Relationships: How Social Contexts Affect the Emergence and Discontinuation of Relationships." *Social Networks* 37 (1): 65–80.

Monge, Peter, and Noshir Contractor. 2003. *Theories of Communication Networks.* New York: Oxford University Press.

Monti, Daniel. 1994. *Wannabe: Gangs in Suburbs and Schools.* Cambridge, Massachusetts: Blackwell.

Moody, James. 2001. "Race, School Integration, and Friendship Segregation in America." *American Journal of Sociology* 107 (3): 679–716.

Moody, James. 2009. "Network Dynamics." In: *The Oxford Handbook of Analytical Sociology*, edited by Peter Bearman and Peter Hedström, 447–474. New York: Oxford University Press.

Moody, James, Daniel McFarland, and Skye Bender-deMoll. 2005. "Dynamic Network Visualization." *American Journal of Sociology* 110 (4): 1206–1241.

Moody, James, and Douglas White. 2003. "Structural Cohesion and Embeddedness: A Hierarchical Concept of Social Groups." *American Sociological Review* 68 (1): 103–127.

Moore, Gwen. 1990. "Structural Determinants of Men's and Women's Networks." *American Sociological Review* 55 (5): 726–735.

Morgner, Christian, ed. 2020. *John Dewey and the Notion of Trans-Action*. New York: Palgrave Macmillan.

Morrill, Calvin, David Snow, and Cynthia White, eds. 2005. *Together Alone: Personal Relationships in Public Places*. Berkeley: University of California Press.

Morris, Charles. 1971. *Writings on the General Theory of Signs*. Den Haag: Mouton.

Mouw, Ted, and Barbara Entwisle. 2006. "Residential Segregation and Interracial Friendship in Schools." *American Journal of Sociology* 112 (2): 394–441.

Mullins, Nicholas. 1973. *Theories and Theory Groups in Contemporary American Sociology*. New York: Harper & Row.

Mützel, Sophie. 2002. *Making Meaning of the Move of the German Capital: Networks, Logics, and the Emergence of Capital City Journalism*. Ann Arbor: UMI.

Mützel, Sophie. 2009. "Networks as Culturally Constituted Processes: A Comparison of Relational Sociology and Actor-Network Theory." *Current Sociology* 57 (6): 871–887.

Nadel, Siegfried. 1957. *The Theory of Social Structure*. London: Cohen & West.

Newman, Mark, and Michelle Girvan. 2004. "Finding and Evaluating Community Structure in Networks." *Physical Review E* 69 (2): 026113.

Niccol, Andrew. 1998. *The Truman Show*. Los Angeles: Paramount Pictures.

Nisbet, Robert. [1953] 1969. *The Quest for Community*. New York: Oxford University Press.

Okamoto, Dina, and Cristina Mora. 2014. "Panethnicity." *Annual Review of Sociology* 40: 219–239.

Opp, Karl-Dieter, and Christiane Gern. 1993. "Dissident Groups, Personal Networks, and Spontaneous Cooperation: The East German Revolution of 1989." *American Sociological Review* 58 (5): 659–680.

Quillian, Lincoln, and Mary Campbell. 2003. "Beyond Black and White: The Present and Future of Multiracial Friendship Segregation." *American Sociological Review* 68 (4): 540–566.

Pachucki, Mark, and Ronald Breiger. 2010. "Cultural Holes: Beyond Relationality in Social Networks and Culture." *Annual Review of Sociology* 36: 205–224.

Padgett, John. 2012a. "From Chemical to Social Networks." In: John Padgett and Walter Powell, *The Emergence of Organizations and Markets*, 92–114. Princeton: Princeton University Press.

Padgett, John. 2012b. "Autocatalysis in Chemistry and the Origin of Life." In: John Padgett and Walter Powell: *The Emergence of Organizations and Markets*, 33–69. Princeton: Princeton University Press.

Padgett, John, and Christopher Ansell. 1993. "Robust Action and the Rise of the Medici." *American Journal of Sociology* 98 (6): 1259–1319.

Padgett, John, and Paul McLean. 2006. "Organizational Invention and Elite Transformation: The Birth of Partnership Systems in Renaissance Florence." *American Journal of Sociology* 111 (5): 1463–1568.

Padgett, John, and Walter Powell. 2012a. *The Emergence of Organizations and Markets*. Princeton: Princeton University Press.

Padgett, John, and Walter Powell. 2012b. "The Problem of Emergence." In: John Padgett and Walter Powell: *The Emergence of Organizations and Markets*, 1–29. Princeton: Princeton University Press.

Paik, Anthony, and Kenneth Sanchagrin. 2013. "Social Isolation in America: An Artifact." *American Sociological Review* 78 (3): 339–360.

Papachristos, Andrew. 2009. "Murder by Structure: Dominance Relations and the Social Structure of Gang Homicide." *American Journal of Sociology* 115 (1): 74–128.

Papachristos, Andrew. 2013. "The Importance of Cohesion for Gang Research, Policy, and Practice." *Criminology & Public Policy* 12 (1): 49–58.

Papilloud, Christian, and Eva-Maria Schultze. 2018. "Pierre Bourdieu and Relational Sociology." In: *The Palgrave Handbook of Relational Sociology*, edited by François Dépelteau, 343–356. New York: Palgrave.

Pappi, Franz Urban, and Peter Kappelhoff. 1984. "Abhängigkeit, Tausch und kollektive Entscheidung in einer Gemeindeelite." *Zeitschrift für Soziologie* 13 (2): 87–117.

Park, Robert. 1950. *Race and Culture*. New York: Free Press.

Park, Robert, and Ernest Burgess. [1921] 1969. *Introduction to the Science of Sociology*. Chicago: University of Chicago Press.

Park, Robert, and Ernest Burgess. [1925] 1967. *The City*. Chicago: University of Chicago Press.

Parsons, Talcott. [1943] 1954. "The Kinship System of the Contemporary United States." In: Talcott Parsons: *Essays in Sociological Theory*, 177–196. New York: The Free Press.

Parsons, Talcott. [1951] 1964. *The Social System*. London: Routledge.

Parsons, Talcott. 1961. "An Outline of the Social System." In: *Theories of Society*, edited by Talcott Parsons, Edward Shils, Kaspar Naegele, and Jesse Pitts, 30–79. Glencoe: Free Press.

Parsons, Talcott. [1968] 1977. "Social Interaction." In: Talcott Parsons: *Social Systems and the Evolution of Action Theory*, 154–176. New York, Free Press.

Parsons, Talcott. [1974] 1977. "On Building Social System Theory: A Personal History." In: Talcott Parsons: *Social Systems and the Evolution of Action Theory*, 22–76. New York: Free Press.

Parsons, Talcott, Edward Shils, Gordon Allport, Clyde Kluckhohn, Henry Murray, Robert Sears, Richard Sheldon, Samuel Stouffer, and Edward Tolman. [1951] 1959. "Some Fundamental Categories of the Theory of Action: A General Statement." In: *Toward a General Theory of Action*, edited by Talcott Parsons and Edward Shils, 3–29. Cambridge, Massachusetts: Harvard University Press.

Pedahzur, Ami, and Arie Perliger. 2006. "The Changing Nature of Suicide Attacks: A Social Network Perspective." *Social Forces* 84 (4): 1987–2008.

Pickering, Paula. 2006. "Generating Social Capital for Bridging Ethnic Divisions in the Balkans: Case Studies of Two Bosniak Cities." *Ethnic and Racial Studies* 29 (1): 79–103.

Plessner, Helmuth. [1960] 1985. "Soziale Rolle und menschliche Natur." In: Helmuth Plessner: *Gesammelte Schriften X*, 227–240. Frankfurt am Main: Suhrkamp.

Plessner, Helmuth. 1969. "De homine abscondito." *Social Research* 36 (4): 497–509.

Popitz, Heinrich. [1967] 1972. "The Concept of Social Role as an Element of Sociological Theory." In: *Role*, edited by John Jackson, 11–39. Cambridge: Cambridge University Press.

Popitz, Heinrich. 2006. *Soziale Normen*. Frankfurt am Main: Suhrkamp.

Popper, Karl. [1935] 2002. *The Logic of Scientific Discovery*. London: Routledge.

Popper, Karl. [1963] 2002. *Conjectures and Refutations*. London: Routledge.

Portes, Alejandro. 1998. "Social Capital: Its Origins and Applications in Modern Sociology." *Annual Review of Sociology* 24: 1–24.

Portes, Alejandro. 2010. *Economic Sociology*. Princeton: Princeton University Press.

Portes, Alejandro, and Julia Sensenbrenner. 1993. "Embeddedness and Immigration: Notes on the Social Determinants of Economic Action." *American Journal of Sociology* 98 (6): 1320–1350.

Powell, Christopher. 2013. "Radical Relationism: A Proposal." In: *Conceptualizing Relational Sociology*, edited by Christopher Powell and François Dépelteau, 187–207. New York: Palgrave Macmillan.

Powell, Christopher, and François Dépelteau, eds. 2013. *Conceptualizing Relational Sociology*. New York: Palgrave Macmillan.

Powell, Walter, Douglas White, Kenneth Koput, and Jason Owen-Smith. 2005. "Network Dynamics and Field Evolution: The Growth of Inter-Organizational Collaboration in the Life Sciences." *American Journal of Sociology* 110 (4): 1132–1205.

Prandini, Riccardo. 2015. "Relational Sociology: A Well-Defined Sociological Paradigm or a Challenging 'Relational Turn' in Sociology?" *International Review of Sociology* 25 (1): 1–14.

Putnam, Hilary. 1981. *Reason, Truth and History*. Cambridge: Cambridge University Press.

Quine, W. V. 1951. "Two Dogmas of Empiricism." *The Philosophical Review* 60 (1): 20–43.

Radcliffe-Brown, A.R. 1940. "On Social Structure." *The Journal of the Royal Anthropological Institute of Great Britain and Ireland* 70 (1): 1–12.

Rainie, Lee, and Barry Wellman. 2012. *Networked*. Cambridge, Massachusetts: MIT Press.

Raymond, Geoffrey. 2003. "Grammar and Social Organization: Yes/No Interrogatives and the Structure of Responding." *American Sociological Review* 68 (6): 939–967.

Raymond, Geoffrey, and John Heritage. 2006. "The Epistemics of Social Relations: Owning Grandchildren." *Language in Society* 35 (5): 677–705.

Reckwitz, Andreas. 2002. "Toward a Theory of Social Practices: A Development in Culturalist Theorizing." *European Journal of Social Theory* 5 (2): 243–263.

Reed, Isaac Ariail. 2011. *Interpretation and Social Knowledge: On the Use of Theory in the Human Sciences*. Chicago: University of Chicago Press.

Rehberg, Karl-Siegbert. 1994. "Rollen als symbolische Ordnungen." In: *Die Eigenart der Institutionen*, edited by Gerhard Göhler, 47–84. Baden-Baden: Nomos.

Ridgeway, Cecilia. 1991. "The Social Construction of Status Value: Gender and Other Nominal Characteristics." *Social Forces* 70 (2): 367–386.

Ridgeway, Cecilia. 2009. "Framed Before We Know It: How Gender Shapes Social Relations." *Gender & Society* 23 (2): 145–160.

Ridgeway, Cecilia, and Shelley Correll. 2004. "Unpacking the Gender System: A Theoretical Perspective on Gender Beliefs and Social Relations." *Gender & Society* 18 (4): 510–531.

Ridgeway, Cecilia, and Lynn Smith-Lovin. 1999. "The Gender System and Interaction." *Annual Review of Sociology* 25: 191–216.

Riessman, Catherine Kohler. 1990. *Divorce Talk*. New Brunswick: Rutgers University Press.

Rivera, Mark, Sara Soderstrom, and Brian Uzzi. 2010. "Dynamics of Dyads in Social Networks: Assortative, Relational, and Proximity Mechanisms." *Annual Review of Sociology* 36: 91–115.

Rorty, Richard. [1979] 2009. *Philosophy and the Mirror of Nature*. Princeton: Princeton University Press.

Rosenfeld, Michael. 2008. "Racial, Educational, and Religious Endogamy in Comparative Historical Perspective." *Social Forces* 87 (1): 1–32.

Roth, Camille, and Jean-Philippe Cointet. 2010. "Social and Semantic Coevolution in Knowledge Networks." *Social Networks* 32 (1): 16–29.

Rubin, Gayle. 1975. "The Traffic in Women: Notes on the 'Political Economy' of Sex." In: *Toward an Anthropology of Women*, edited by Rayna Reiter, 157–210. New York: Monthly View Press.

Ruggieri, Davide. 2014. "Fuhse and Donati on Relational Sociology: Beyond the Structural View of Social Networks." *Sociologia e Politiche Sociali* 17 (1): 51–70.

Rule, Alix, Jean-Philippe Cointet, and Peter Bearman. 2015. "Lexical Shifts, Substantive Changes, and Continuity in State of the Union Discourse, 1790–2014." *Proceedings of the National Academy of Sciences* 112 (35): 10837–10844.

Rytina, Steve, and David Morgan. 1982. "The Arithmetic of Social Relations: The Interplay of Category and Network." *American Journal of Sociology* 88 (1): 88–113.

Sacks, Harvey, Emanuel Schegloff, and Gail Jefferson. 1974. "A Simplest Systematics for the Organization of Turn-Taking for Conversation." *Language* 50 (4): 696–735.

Sageman, Mark. 2011. *Understanding Terror Networks*. Philadelphia: University of Pennsylvania Press.

Salvini, Andrea. 2010. "Symbolic Interactionism and Social Network Analysis." *Symbolic Interaction* 33 (3): 364–388.

Schatzki, Theodore. 1996. *Social Practices*. Cambridge: Cambridge University Press.

Schegloff, Emanuel. 2007. *Sequence Organization in Interaction*. New York: Cambridge University Press.

Schegloff, Emanuel, and Harvey Sacks [1973] 1984. "Opening Up Closings." In: *Language in Use*, edited by John Baugh and Joel Sherzer, 69–99. Englewood Cliffs: Prentice-Hall.

Schelsky, Helmut. 1970. "Zur soziologischen Theorie der Institution." In: *Zur Theorie der Institution*, edited by Helmut Schelsky, 10–26. Düsseldorf: Bertelsmann.

Schiffrin, Deborah. 1987. *Discourse Markers*. Cambridge: Cambridge University Press.

Schiffrin, Deborah. 1994. *Approaches to Discourse*. Oxford: Blackwell.

Schmitt, Marco. 2009. *Trennen und Verbinden: Soziologische Untersuchungen zur Theorie des Gedächtnisses*. Wiesbaden: VS.

Schmitt, Marco, and Jan Fuhse. 2015. *Zur Aktualität von Harrison White*. Wiesbaden: VS.

Schneider, Wolfgang Ludwig. 2000. "The Sequential Production of Social Acts in Conversation." *Human Studies* 23 (2): 123–144.

Schütz, Alfred. [1932] 1967. *The Phenomenology of the Social World*. Evanston: Northeastern University Press.

Schütz, Alfred. [1944] 1971. "The Stranger: An Essay in Social Psychology." In: Alfred Schütz: *Collected Papers II*, 91–105. Den Haag: Nijhoff.

Schweizer, Thomas. 1996. *Muster sozialer Ordnung: Netzwerkanalyse als Fundament der Sozialethnologie*. Berlin: Reimer.

Schweizer, Thomas, Michael Schnegg, and Susanne Berzborn. 1998. "Personal Networks and Social Support in a Multiethnic Community in Southern California." *Social Networks* 20 (1): 1–21.

Schweizer, Thomas, and Douglas White, eds. 1998. *Kinship, Networks, and Exchange*. New York: Cambridge University Press.

Scott, John. 2000. *Social Network Analysis*. 2nd edition. London: Sage.

Seeley, Lotus. 2014. "Harrison White as (Not Quite) Poststructuralist." *Sociological Theory* 32 (1): 27–42.

Shannon, Claude, and Warren Weaver. 1949. *The Mathematical Theory of Communication*. Urbana: University of Illinois Press.

Sharma, Ursula. 1999. *Caste*. Buckingham: Open University Press.

Sherif, Muzafer. 1966. *In Common Predicament*. Boston: Houghton Mifflin.

Shibutani, Tamotsu. 1955. "Reference Groups as Perspectives." *American Journal of Sociology* 60 (6): 562–569.

Shibutani, Tamotsum, and Kian Kwan. 1965. *Ethnic Stratification*. New York: Macmillan.

Shils, Edward. 1951. "The Study of the Primary Group." in: *The Policy Sciences*, edited by Daniel Lerner and Harold Lasswell, 44–69. Stanford: Stanford University Press.

Shils, Edward. 1962. "The Theory of Mass Society." *Diogenes* 10 (39): 45–66.

Shrum, Wesley, Neil Cheek, and Saundra Hunter. 1988. "Friendship in School: Gender and Racial Homophily." *Sociology of Education* 61 (4): 227–239.

Silver, Allan. 1990. "Friendship in Commercial Society: Eighteenth-Century Social Theory and Modern Sociology." *American Journal of Sociology* 95 (6): 1474–1504.

Simmel, Georg. [1908] 1950. *The Sociology of Georg Simmel*. New York: The Free Press.

Simmel, Georg. [1908] 1964. *Conflict & The Web of Group-Affiliations*. New York: Free Press.

Simmel, Georg. [1908] 1971. "The Stranger." In: *Georg Simmel: On Individuality and Social Forms*, 143–149. Chicago: University of Chicago Press.

Simmel, Georg. 1909. "The Problem of Sociology." *American Journal of Sociology* 15 (3): 289–320.

Singh, Sourabh. 2016. "What Is Relational Structure? Introducing History to the Debates on the Relation Between Fields and Social Networks." *Sociological Theory* 34 (2): 128–150.

Small, Mario. 2017. *Someone to Talk to*. New York: Oxford University Press.

Smith, Christian. 2010. *What Is a Person? Rethinking Humanity, Social Life, and the Moral Good from the Person Up*. Chicago: University of Chicago Press.

Smith, Jeffrey, Miller McPherson, and Lynn Smith-Lovin. 2014. "Social Distance in the United States: Sex, Race, Religion, Age, and Education Homophily Among Confidants, 1985 to 2004." *American Sociological Review* 79 (3): 432–456.

Smith, Sanne, Ineke Maas, and Frank van Tubergen. 2014. "Ethnic Ingroup Friendships in Schools: Testing the By-Product Hypothesis in England, Germany, the Netherlands and Sweden." *Social Networks* 39 (1): 33–45.

Smith, Sanne, Frank van Tubergen, Ineke Maas, and Daniel McFarland. 2016. "Ethnic Composition and Friendship Segregation: Differential Effects for Adolescent Natives and Immigrants." *American Journal of Sociology* 121 (4): 1223–1272.

Smith, Tammy. 2007. "Narrative Boundaries and the Dynamics of Ethnic Conflict and Conciliation." *Poetics* 35 (1): 22–46.

Smith-Lovin, Lynn, and Charles Brody. 1989. "Interruptions in Group Discussions: The Effects of Gender and Group Composition." *American Sociological Review* 54 (3): 424–435.

Smith-Lovin, Lynn, and Miller McPherson. 1993. "You Are Who You Know: A Network Approach to Gender." In: *Theory on Gender / Feminism on Theory*, edited by Paula England, 223–251. New York: de Gruyter.

Snijders, Tom. 2011. "Statistical Models for Social Networks." *Annual Review of Sociology* 37: 131–153.

Snijders, Tom, Gerhard van de Bunt, and Christian Steglich. 2010. "Introduction to Stochastic Actor-Based Models for Network Dynamics." *Social Networks* 32 (1): 44–60.

Snow, David, Burke Rochford, Steven Worden, and Robert Benford. 1986. "Frame Alignment Processes, Micromobilization, and Movement Participation." *American Sociological Review* 51 (4): 464–481.

Somers, Margaret. 1994. "The Narrative Constitution of Identity: A Relational and Network Approach." *Theory and Society* 23 (5): 605–649.

Spencer Brown, George. [1969] 1972. *Laws of Form*. New York: Julian Press.

Spencer-Oatey, Helen. 2002. "Managing Rapport in Talk: Using Rapport Sensitive Incidents to Explore the Motivational Concerns Underlying the Management of Relations." *Journal of Pragmatics* 34 (5): 529–545.

Srinivas, M.N., and André Béteille. 1964. "Networks in Indian Social Structure." *Man* 64 (212): 165–168.

Stebbins, Robert. 1969. "Social Network as a Subjective Construct: A New Application for an Old Idea." *Canadian Review of Sociology and Anthropology* 6 (1): 1–14.

Stevenson, Nick. 2002. *Understanding Media Cultures*. 2nd edition. London: SAGE.

Stinchcombe, Arthur. 1968. *Constructing Social Theories*. New York: Harcourt, Brace & World.

Stinchcombe, Arthur. 2005. *The Logic of Social Research*. Chicago: University of Chicago Press.

Sudman, Seymour, Norman Bradburn, and Norbert Schwarz. 1996. *Thinking About Answers*. San Francisco: Jossey-Bass.

Sumner, William Graham. [1906] 1959. *Folkways*. New York: Dover.

Swedberg, Richard. 2014. *The Art of Social Theory*. Princeton: Princeton University Press.

Swedberg, Richard. 2017. "Theorizing in Sociological Research: A New Perspective, a New Departure." *Annual Review of Sociology* 43: 189–206.

Swidler, Ann. 1986. "Culture in Action: Symbols and Strategies." *American Sociological Review* 51 (2): 273–286.

Swidler, Ann. 2001. *Talk of Love: How Culture Matters*. Chicago: University of Chicago Press.

Tacke, Veronika. 2000. "Netzwerk und Adresse." *Soziale Systeme* 6 (2): 291–320.

Tajfel, Henri. 1981. *Human Groups and Social Categories*. Cambridge: Cambridge University Press.

Tajfel, Henri, ed. 1982. *Social Identity and Intergroup Identifications*. Cambridge: Cambridge University Press.

Tannen, Deborah. [1990] 2007. *You Just Don't Understand: Women and Men in Conversation*. New York: Harper.

Tannen, Deborah. 1993. "What's in a Frame? Surface Evidence for Underlying Expectations." In: *Framing in Discourse*, edited by Deborah Tannen, 14–56. New York: Oxford University Press.

Tannen, Deborah. 1996. *Gender and Discourse*. New York: Oxford University Press.

Taylor, Charles. 1989. *Sources of the Self*. Cambridge: Cambridge University Press.

Thomas, William. [1927] 1966. "Situational Analysis: The Behavior Pattern and the Situation." In: William Thomas, *On Social Organization and Social Personality*, 154–167. Chicago: University of Chicago Press.

Tilly, Charles. 1998. *Durable Inequality*. Berkeley: University of California Press.

Tilly, Charles. 2002. *Stories, Identities, and Political Change*. Lanham: Rowman & Littlefield.

Tilly, Charles. 2003. *The Politics of Collective Violence*. New York: Cambridge University Press.

Tilly, Charles. 2005a. *Identities, Boundaries and Social Ties*. Boulder: Paradigm.

Tilly, Charles. 2005b. *Trust and Rule*, New York: Cambridge University Press.

Tilly, Charles. 2006. *Why?* Princeton: Princeton University Press.

Tilly, Charles. 2008. *Explaining Social Processes*. Boulder: Paradigm.

Tilly, Charles, and Leslie Wood. 2009. *Social Movements 1768–2008*. Boulder: Paradigm.

Tuomela, Raimo, and Kaarlo Miller. 1988. "We-Intentions." *Philosophical Studies* 53 (3): 367–389.

Turner, Ralph. 1962. "Role-Taking: Process Versus Conformity." In: *Human Behavior and Social Process*, edited by Arnold Rose, 20–40. Boston: Houghton Mifflin.

Turner, Stephen. 1994. *The Social Theory of Practices*. Chicago: University of Chicago Press.

van Atteveldt, Wouter, Tamir Sheafer, Shaul Shenhav, and Yair Fogel-Dror. 2017. "Clause Analysis: Using Syntactic Information to Automatically Extract Source, Subject, and Predicate from Texts with an Application to the 2008–2009 Gaza War." *Political Analysis* 25 (2): 207–222.

van Dijk, Jan. 2012. *The Network Society*. 3rd edition. London: Sage.

van Fraassen, Bas. 1980. *The Scientific Image*. Oxford: Clarendon Press.

Verbrugge, Lois. 1977. "The Structure of Adult Friendship Choices." *Social Forces* 56 (2): 576–597.

von Wiese, Leopold. 1932. *Systematic Sociology: On the Basis of the Beziehungslehre and Gebildelehre*. New York: Wiley.

Wasserman, Stanley, and Katherine Faust. 1994. *Social Network Analysis*. New York: Cambridge University Press.

Watts, Duncan. 1999. "Networks, Dynamics, and the Small-World Phenomenon." *American Journal of Sociology* 105 (2): 493–527.

Watzlawick, Paul, Janet Beavin, and Don Jackson. 1967. *Pragmatics of Human Communication*. New York: Norton.

Weber, Max. [1913] 1981. "Some Categories of Interpretive Sociology." *The Sociological Quarterly* 22 (2): 151–180.

Weber, Max. [1922] 1978. *Economy and Society*. Berkeley: University of California Press.

Wellman, Barry. 1983. "Network Analysis: Some Basic Principles." *Sociological Theory* 1: 155–200.

Wellman, Barry, and Stephen Berkowitz, eds. 1988. *Social Structures: A Network Approach*. New York: Cambridge University Press.

West, Candace, and Sarah Fenstermaker. 1995. "Doing Difference." *Gender and Society* 9 (1): 8–37.

West, Candace, and Don Zimmerman. 1987. "Doing Gender." *Gender and Society* 1 (2): 125–151.

White, Douglas, Jason Owen-Smith, James Moody, and Walter Powell. 2004. "Networks, Fields and Organizations." *Computational & Mathematical Organization Theory* 10 (1): 95–117.

White, Harrison. 1963. *An Anatomy of Kinship*. Englewood Cliffs: Prentice-Hall.

White, Harrison. [1965] 2008. "Notes on the Constituents of Social Structure." *Sociologica* 2008 (1): doi: 10.2383/26576.

White, Harrison. 1992. *Identity and Control: Towards a Structural Theory of Action*. Princeton: Princeton University Press.

White, Harrison. 1993. *Careers & Creativity: Social Forces in the Arts*. Boulder: Westview.

White, Harrison. 1995a. "Network Switchings and Bayesian Forks: Reconstructing the Social and Behavioral Sciences." *Social Research* 62 (4): 1035–1063.

White, Harrison. 1995b. "Passages réticulaires, acteurs et grammaire de la domination." *Revue française de sociologie* 36 (4): 705–723.

White, Harrison. 2008. *Identity and Control: How Social Formations Emerge*. Princeton: Princeton University Press.

White, Harrison, Scott Boorman, and Ronald Breiger. 1976. "Social Structure from Multiple Networks. I. Blockmodels of Roles and Positions." *American Journal of Sociology* 81 (4): 730–780.

White, Harrison, and Ronald Breiger. 1976. "Pattern Across Networks." *Society* 12 (5): 68–74.

White, Harrison, Jan Fuhse, Matthias Thiemann, and Larissa Buchholz. 2007. "Networks and Meaning: Styles and Switchings." *Soziale Systeme* 13 (1+2): 534–555.

White, Robert. 1989. "From Peaceful Protest to Guerilla War: Micromobilizations of the Provisional Irish Republican Army." *American Journal of Sociology* 94 (6): 1277–1302.

Whitehead, Alfred North. [1929] 1978. *Process and Reality*. New York: The Free Press.

Willer, David, Henry Walker, Barry Markovsky, Robb Willer, Michael Lovaglia, Shane Thye, and Brent Simpson. 2002. "Network Exchange Theory." In: *New Directions in Contemporary Sociological Theory*, edited by Joseph Berger and Morris Zelditch, 109–144. Lanham: Rowman & Littlefield.

Wimmer, Andreas. 2013. *Ethnic Boundary Making*. New York: Oxford University Press.

Wimmer, Andreas. 2015. "Race-Centrism: A Critique and a Research Agenda." *Ethnic and Racial Studies* 38 (13): 2186–2205.

Wimmer, Andreas, and Kevin Lewis. 2010. "Beyond and Below Racial Homophily: ERG Models of Friendship Network Documented on Facebook." *American Journal of Sociology* 116 (2): 583–642.

Winship, Christopher, and Michael Mandel. 1983. "Roles and Positions: A Critique and Extension of the Blockmodeling Approach." *Sociological Methodology* 14: 314–344.

Wittig, Monique. 1980. "The Straight Mind." *Feminist Issues* 1 (1): 104–111.

Wittig, Monique. 1982. "The Category of Sex." *Feminist Issues* 2 (2): 64–68.

Wolf, Eric. 1966. "Kinship, Friendship, and Patron-Client Relations in Complex Societies." In: *The Social Anthropology of Complex Societies*, edited by Michael Banton, 1–22. London: Tavistock.

Wolff, Kurt. 1950. "Introduction." In: *Georg Simmel: The Sociology of Georg Simmel*, edited and translated by Kurt Wolff, xvii–lxiv. New York: The Free Press.

Wood, Julia. 1982. "Communication and Relational Culture: Bases for the Study of Human Relationships." *Communication Quarterly* 30 (2): 75–83.

Wood, Julia, and Steve Duck, eds. 2006. *Composing Relationships*. Belmont: Wadsworth.

Wouters, Cas. 1998. "Balancing Sex and Love Since the 1960s Sexual Revolution." *Theory, Culture & Society* 15 (3–4): 187–214.

Wright, Erik Olin, and Donmoon Cho. 1992. "The Relative Permeability of Class Boundaries to Cross-Class Friendships." *American Sociological Review* 57 (1): 85–102.

Wrong, Dennis. 1961. "The Oversocialized Conception of Man in Modern Sociology." *American Sociological Review* 26 (2): 183–193.

Wuthnow, Robert. 1989. *Meaning and Moral Order*. Berkeley: University of California Press.

Yablonski, Lewis. 1959. "The Delinquent Gang as a Near-Group." *Social Problems* 7 (2): 108–117.

Yeung, King-To. 2005. "What Does Love Mean? Exploring Network Culture in Two Network Settings." *Social Forces* 84 (1): 391–420.

Yeung, King-To, and John Levi Martin. 2003. "The Looking Glass Self: An Empirical Test and Elaboration." *Social Forces* 81 (3): 843–879.

Yinger, Milton. 1985. "Ethnicity." *Annual Review of Sociology* 11: 151–180.

Zablocki, Benjamin. 1980. *Alienation and Charisma: A Study of Contemporary American Communes*. New York: The Free Press.

Zelizer, Viviana. 2005. *The Purchase of Intimacy*. Princeton: Princeton University Press.

Zerubavel, Eviatar. 1991. *The Fine Line: Making Distinctions in Everyday Life*. Chicago: University of Chicago Press.

Znaniecki, Florian. 1939. "Social Groups as Products of Participating Individuals." *American Journal of Sociology* 44 (6): 799–811.

Znaniecki, Florian. 1954. "Social Groups in the Modern World." In: *Freedom and Control in Modern Society*, edited by Morroe Berger, Theodore Abel, and Charles Page, 125–140. Toronto: Van Nostrand.

Index